COLLECTING MECHANICAL ANTIQUES

Other Books by Ronald Pearsall

The Worm in the Bud (Weidenfeld & Nicolson)
The Table-Rappers (Michael Joseph)
Victorian Sheet Music Covers (David & Charles)

In Preparation

Collecting Scientific Instruments (David & Charles)

COLLECTING MECHANICAL ANTIQUES

Ronald Pearsall

ARCO PUBLISHING COMPANY, INC
New York

Published by Arco Publishing Company, Inc.
219 Park Avenue South, New York, N.Y. 10003

Copyright © 1973 by Ronald Pearsall

All rights reserved

Library of Congress Catalog Card Number 73-76415

ISBN 0-668-02967-6

Printed in Great Britain

CONTENTS

LIST OF ILLUSTRATIONS

INTRODUCTION

THE TWO conditions for entry in COLLECTING MECH-
ANICAL ANTIQUES are that the object is interesting in itself,
and that it is portable. It has also been necessary to differentiate
between a mechanical antique and an instrument. An instrument
is defined in the *Oxford English Dictionary* as being a tool,
implement, or weapon, and is derived from *instruere*, to instruct.
An instrument measures or informs, a tool, interesting though it
may be, makes. A mechanical antique is used.

It is used either in industry or the home, either as an aid or
a substitute for human labour, or for entertainment. A mechanical antique in the sense used throughout this book is not a
manufacturer; whether activated by electricity, simple gearing, or
clockwork, it processes rather than makes, and perhaps the only
item where more comes out than goes in is the vacuum cleaner!

Mechanical antiques are the end products of a manufacturing
process, and operate by using natural phenomena, such as electricity (the telephone or the telegraph), by working upon material
inserted into the object (the typewriter or the sewing machine),
or by being programmed to achieve a certain effect (automata
or musical boxes).

No hard and fast rules have been laid down about dates,
though it will be found that a large proportion of the items first
saw production (rather than inception) between 1830 and 1880,
and of these objects most will have originated in America where
the pressures that thwarted mechanisation in Europe did not
operate. In Britain and the Continent there was no desperate need

for labour-saving devices, for there were ample supplies of expert and cheap labour, and only when there was a compelling case for developing an aid to an existing industry did Europeans stir themselves. Necessity was the mother of invention in the electrical telegraph, hurried into use by the demand of the burgeoning railways for instant communication, and in this field Britain was in the van.

Why was America so predominant in the production of mechanical antiques, how is it that she could produce the first commercially successful typewriter when the credit could have gone to Britain, France, or Italy, all working on the problem of the writing machine? Simply that in America labour was scarce and expensive. In countries where there were thousands of clerks painstakingly writing their copperplate hand for a few pounds a year there was no economic need to relieve them of this chore. It was the same with the sewing machine. There were needlewomen in abundance for any amount of work.

It was another story when it came to entertainment. If the promotion and manufacture of typewriters and sewing machines had been as interesting to the upper classes as being held spellbound by mechanical toys, then aids to industry and in the home would have been pursued with a good deal more diligence. The tastes and the inclinations of the dominant classes influenced the progress of mechanisation. Precision engineering that could have gone into the business of making life easier for those at the bottom of the social scale was primarily at the service of watch and clockmaking, and the making of firearms. The taste of the upper classes for automata and musical boxes encouraged the watchmakers to cultivate this field.

It is easy now to appreciate the enormous effect typewriters and telephones have had on the life of everyone, but when they first came on to the market they were novelties with limited possibilities. The telephone was first of all seen as an instrument to connect one room with another. The gramophone was considered purely a business accessory that would record speech. The camera was an ingenious device that would assist artists.

This may have been lack of imagination on the part of the public, but when one looks at some of the primitive mechanisms one cannot disguise the fact that there were ample grounds for honest scepticism.

Inventors throughout the nineteenth century were given a boost by the success of the telegraph; its utility transcended all expectations. It was the first practical use of electricity, until then a scientific amusement. Many far-sighted men were handicapped by the relatively slow progress made in harnessing electricity, and it was not until the end of the century that electric motors were commercially produced, and the labour saving devices in the modern home that we take so much for granted were possible. The vacuum cleaner, the refrigerator, the washing machine, the electric kettle—although these were visualised by the nineteenth century they had to await the present century for not just the electric motor, but for mains electricity.

For collectors with a limited budget, mechanical antiques offer a fascinating and unexploited challenge. They relate not only to their present day successors, but also to their time, and it is no accident that the first telegraph equipment is redolent of Regency craftmanship—the date, after all, was 1837. Nor is it surprising that the first sewing machines are lavished with painted flowers; no decorative artist could resist a black japanned finish, whether or not it was on a sewing machine. The telephones of the turn of the century, gilded with art nouveau motifs, are decorative items in themselves (and can be converted to modern use). The cameras of the second half of the nineteenth century, festooned with brass, are as much the product of the cabinet-maker as the optical worker.

There are one or two things to remember when buying mechanical antiques—certain items may not be as old as they seem. Phonographs were made well into the 1920s, and the gramophones with gigantic horns are by no means antediluvian—the large horns were directly related to the invention of electrical recording in the early 1920s, in which the ideal was a horn at least nine feet long (the Americans were more prosaic, and preferred the horn folded back on itself and hidden in a cabinet).

Electrical prototypes are always worth looking for. They may be the work of some unknown contemporary of Edison striving to solve the problem of the talking machine in his own way. So too are the various gadgets using electricity in ways weird and wonderful; the most common of these are the disregarded and neglected galvanic machines, which gave a mild shock to the user, deliberately and therapeutically.

With the exception of automata, introduced in this book because to omit them would be unthinkable, and certain providers of mechanical music such as the musical box, the items dealt with are available and inexpensive. Even decidedly collectible objects such as phonographs are within the pocket of most collectors. Most of the items were made to last, and not only can they be shown in the drawing room as talking points but they can also be made to work. The sewing machines were produced with a variety of attachments to do the most ingenious things and may well vie with modern machines in versatility. Nineteenth-century typewriters are virtually indestructible, and their honest-to-goodness type-faces may prove to be a quaint alternative to their modern successors. There is nothing better than a horned gramophone to play back period 78 rpm records.

In a time of possible power cuts, when the electric and electronic age might well have not existed, it may be that money spent on pre-electric technological marvels will reap extra dividends!

CHAPTER I

THE SEWING MACHINE

THE SEWING machine was one of the most important inventions of the nineteenth century. It was the first domestic mechanical aid to be mass-produced, and it had an incalculable effect on nineteenth-century life, providing the means, for the first time, of manufacturing cheap clothes. The sewing machine was intricate and sophisticated, designed to do a series of precision tasks in a set order, and if one wanted to draw a parallel one might say that it was the industrial equivalent of the watch.

The evolution and production of the sewing machine fitted into a pattern. It was projected in Europe, developed and promoted in the United States, and then produced under licence or emulated in Britain and the continent.

In 1755 Charles Weisenthal produced a new kind of needle for making embroidery stitches; this needle had an eye at the end nearest to the point. In 1790 a British cabinetmaker, Thomas Saint, patented:

> An Entire New Method of Making and Completing Shoes, Boots, Splatterdashes, Clogs, and Other Articles, by Means of Tools and Machines also Invented by Me for that Purpose, and of Certain Compositions of the Nature of Japan or Varnish, which will be very advantageous in many useful Applications.

The patent also contained details of three machines, one of them for 'stitching, quilting, or sewing', and although impractical this machine incorporated features found in the modern sewing

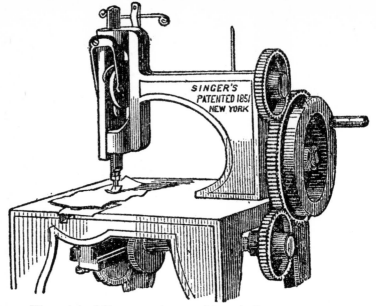

1   The original Singer sewing machine of 1851.

machine—a horizontal cloth plate or table, an overhanging arm carrying a straight needle, and a continuous supply of thread from a spool. The motion was produced by the rotation of a hand crank on a shaft, activating cams. An awl made a hole in the material, a forked needle thrust the thread through the hole. The thread was caught by a looper and kept until enchained in the next loop of thread. There were thread tighteners above and below the material, and an adjustment varied the stitches for different classes of work. Whether Saint actually made the machine is debatable, for when the patent was discovered in 1873 an Englishman tried to mock up this machine, finding in the process that modifications were necessary as the machine projected by Saint would not operate at all. Nevertheless all the ingredients of the sewing machine were there, though the idea was not pursued. Why invent a machine that would take the place of a needlewoman when there were ample supplies of seamstresses willing and eager to do all the machine did at starvation wages?

Those inventors who did venture into this field made the

mistake of trying to reproduce the action of human hands. An 1804 patent utilised pincers imitating the action of fingers. In 1814 Josef Madersperger invented a viable sewing machine in Vienna, but despite the encouragement of the Court nothing came of it. More important was the Thimmonier sewing machine, patented by a French tailor in 1830, which made a chainstitch by means of a barbed needle. The needle thrust through the material, caught a thread from the thread carrier, and brought up a loop from beneath the material. When the process was repeated, the second loop became enchained in the first. The fabric was fed through the stitching mechanism manually, which meant that a constant speed had to be maintained by the operator if the stitches were to be the same length. As this sewing machine

2 *Although Thomas Saint patented his sewing machine in 1790, so far as is known it was never constructed, and this is a mock-up made in 1873 when Saint's patent was rediscovered.*

3 *Thimmonier's sewing machine was the first successful one.*
Patented in 1830, by 1841 Thimmonier had 80 stitching Army
uniforms.

was designed specifically for embroidery and ornamentation rather than utility, the manual feeding of the material was a drawback. But, unlike other pioneers, Thimmonier saw the commercial possibilities of his invention, and by 1841 he had 80 machines stitching army uniforms in a Paris shop. His success was brought to a sharp end by the invasion of the shop by a mob of tailors, out to destroy this threat to their livelihood. They broke up his machines, and Thimmonier fled, but although he later received backing to reorganise his thriving business, the 1848 Revolution prevented this.

In the United States, Walter Hunt was working towards the definitive sewing machine. Hunt was a professional inventor, and over the years he had taken out patents for a knife-sharpener, a stove, an ice boat, a nail machine, an inkwell, a fountain pen, paper collars, and a reversible metal boot-heel. Some time between 1832 and 1834 he evolved a machine that would make a lockstitch; this was the first stitch that did not attempt to imitate a hand stitch. It required two threads, one passing through a loop in the other, both interlocking in the heart of the seam. The combination of an eye-pointed needle with a shuttle containing the second thread was a vital technical breakthrough.

Hunt was not particularly interested in following up the sewing machine, and it was left to Elias Howe, Jr, to correlate all the work of the earlier inventors. In 1846 the modern sewing machine was born, and from then it was only a question of refining and developing. The problem of a regular feed was solved by Allen Wilson, a Michigan cabinetmaker, in 1854. The device used was a horizontally reciprocating toothed surface, with the teeth projecting forward and moving in a rectangular motion. This moved the material forward automatically and continuously, and permitted the turning of the material to produce a curved seam.

It was at this point that the American genius for seeing possibilities came into play. Isaac M. Singer (1811-75) of Pittsdown, New York, patented the first practical domestic sewing machine in 1851. Singer, trained as a mechanic and a cabinetmaker, ironed out the snags and put his machine on the market, using mass-production techniques to make large quantities and using modern methods of advertising to promote it. Although he had been involved in inventing and patenting a novel kind of type-carving machine, he speedily realised the vast potential of the

The ROYAL Anchor.

¶ We respectfully solicit a visit to our newly-enlarged Show Rooms, where will be found A Selection of the Best English and American Family Sewing Machines, which may be seen at work and thoroughly tested before purchase. [Catalogues post free.

THOMAS BRADFORD & CO.,
63, FLEET-STREET, LONDON; and MANCHESTER.

THE "ROYAL ANCHOR" has been proved by thousands of ladies to be the perfection of a Lock-stitch Sewing Machine, and no lady who has applied herself so as readily to produce its most beautiful sample of work, will be satisfied with, or will allow herself to be "advertised" into content with any other kind of stitch, single thread or otherwise—which of course are very simple, and which any child can manage, and for certain ornamental work may be very pleasing and very taking to ladies unacquainted with a good Lock-stitch Machine; but for *profitable use* and *permanent* satisfaction, incontrovertibly proved by the simple fact of its adoption for all manufacturing purposes, we recommend to all ladies a reliable Lock-stitch Machine, of which we believe the "Royal Anchor" to be about, if not the very best (for a lady's use) that is made.

Price complete, with all extras, £5.

4   *A typical sewing machine of 1871.*

sewing machine and dropped all his projects in favour of it. Singer continued to be active in the sewing machine business until 1863, then moved to Paris, and later to Torquay, England, where he conceived the idea of a fabulous Greco-Roman mansion to be named 'The Wigwam'. He died before this was brought to fruition.

The importance of Singer not only in the field of the sewing machine but in the history of technology and marketing is impossible to over-estimate. His sewing machine was the most widely advertised and distributed product of the 1850s, and the industry pioneered instalment purchase (1856) and 'sale and service'. The sewing machine was soon adapted for other purposes, and a massive range of specialised machines was developed for use in such diverse industries as bleaching, carpet, lace, jute, glove, hosiery, hatting, umbrella, corset-making, tent, flag, fur, and particularly the boot and shoe industry. Thomas Saint's ideas of 1790 were bearing fruit.

Sewing machines were soon seen in Britain. In 1854 the periodical *Household Words* marvelled at it:

The iron seamstress is composed of a flat metal surface, about twelve inches square (a very comfortable little body, as it will be seen), resting on four substantial legs. From one side of the lady's flat iron surface, an arm rises to the height of about ten inches, and then, bending the elbow, passes over to the opposite side.

There was some doubt about the morality of inventing such machines:

Will the iron seamstress drive the seamstress of (not much) flesh and blood to more remunerative employments? The answer is not an easy one. Needlework, though poorly paid, has long been the drudgery to which women have taken when the strong arm that shielded them has fallen suddenly away. It was work easily learned and abundantly wanted. Poor creatures whose prospect was so dark that any pittance was a relief, could always, if they would accept the hard price, get the work.

This plaintive little homily puts the finger on why the sewing machine was not pursued with the diligence commensurate with its possibilities in Britain. It would create unemployment and misery for a class that had been publicised and sentimentalised over in Thomas Hood's famous 'Song of a Shirt'.

5   *The first sewing machine in London, made by Howe.*

Impressive as the sewing machine was, would it do all that the needlewoman could do—'those mysteries known to the erudite as flounces, gussets, frills and tucks?' The sceptical thought not. But a reporter of the *Illustrated Magazine of Art* writing at the same time went to see a demonstration of the sewing machine in Cheapside. 'Every minute a yard of cloth is sewn in a style far superior to that of hand-sewing, and the invention can be applied to nearly every sort of work.'

The *Illustrated Magazine of Art* saw the sewing machine as a boon. It would not throw seamstresses out of work and, most likely, on to the streets, but relieve them of their degrading employment. It 'promises to make the wrongs of distressed needlewomen a matter of historical record, instead of a living and painful reality'.

In America, the commercial possibilities of the sewing machine had resulted in a patent war. Entrepreneurs were anxious to break the monopoly of Singer, and began grubbing about among earlier sewing machine patents to see what could be salvaged. To complicate matters, Elias Howe, responsible for the first workable domestic machine but totally lacking the business acumen and genius of Singer, demanded $25,000 from Singer for infringement of his patent. A number of rival manufacturers, seeing a way into the market, purchased licenses from Howe and advertised that anyone could sell their machines without fear of a lawsuit.

Singer was investing large sums of money in marketing the sewing machine, setting up lavishly-decorated showrooms with carved walnut furniture, rich in ormolu and gilt, and deep carpet. The machines were demonstrated by pretty young women. The whole was a new concept of selling, a prototype for all later marketing, whether it was the typewriter in the 1870s or the motor-car in the opening years of this century.

The lawsuits promised and pending seemed to Singer to be mutually destructive, and the possibilities were that all the manufacturers would lose, and only Howe would be the winner —the rights which Howe had sold to the rival manufacturers of Singer brought him in a royalty of $25 on each machine. However, Howe was hoist with his own petard, for in an attempt to improve his own machine he was himself sued.

In 1856 one of the leading manufacturers of sewing machines

advanced the idea of a 'Sewing Machine Combination'. This syndicate was eagerly taken up. The leading manufacturer in terms of output was not, as might be supposed, Singer, but the Wheeler and Wilson Manufacturing Company, who retained a dominance until 1866 when they were overtaken by Singer, who thereafter improved their position. In 1866 Singer produced 30,960 machines; in 1870 127,833; in 1876 262,316.

The sewing machine sold in England in 1854 was not cheap. It cost £30—more than six months pay for a working man. The idea of the hire-purchase system promoted by Singer was anathema to British sensibilities. Attempts were made to bring the price of sewing machines down. This was done by producing enough machines to make price-cutting viable, or by introducing smaller machines. James Gibbs was a pioneer of the cheaper machine, marketing a $50 model in 1859, while equivalent

6 *A replica of Howe's prototype sewing machine. Dating from 1846, Howe was soon overwhelmed by Singer, whom he sued for infringement of patent rights.*

models by other companies cost twice as much. To compete, Singer produced a $50 machine, but this was too light, though its successor, costing $75, was more successful.

Some of the cheap machines produced in the 1850s and 1860s were no more than stunts. Heyer's Pocket Sewing Machine of 1863 was a bent piece of metal strip two inches long and two inches wide. The five dollar 'Fairy' sewing machine was hardly more elaborate, and a copy of this, the 'Gold Medal' of 1862 was advertised:

> A first class sewing machine, handsomely ornamented, with all working parts silver plated. Put up in a highly polished mahogany case, packed ready for shipment. Price $10.00.

Not surprisingly, many purchasers considered that they had been swindled, as they expected a heavy-duty machine.

Although Singer would in future years dominate the sewing machine market both in America and in England, the British version of the sewing machine first seen was not his, but Howe's. Elias Howe had sent his brother to England to sound out this potentially profitable market. At that time England had a population equal to that of the United States and was considerably richer. Amasa Howe was discouraged by the initial lack of interest by the British, not appreciating that the economic situation was different. America was short of skilled labour, and there was no abundance of needlewomen or tailors. Wages were high, the population was sparse, and any device that would increase the productivity of an expensive labour force was sure of a welcome.

Eventually Howe met William Thomas, a manufacturer of umbrellas, corsets, and leather goods, who offered to buy the sewing machine for £250. He undertook to engage Howe at a salary of £3 a week if Howe would adapt this machine to the making of corsets. Elias Howe sailed for London in 1847 taking with him the sewing machine and the patent papers, and later William Thomas advanced sufficient money for Howe's wife and family to join the brothers in England.

During the construction of the second machine for Thomas, Howe's money ran out and he was forced to send his family home. He worked on the machine for several months, but was forced to sell it for five pounds, pawn his precious first machine and the patent papers, and return to America, to find that

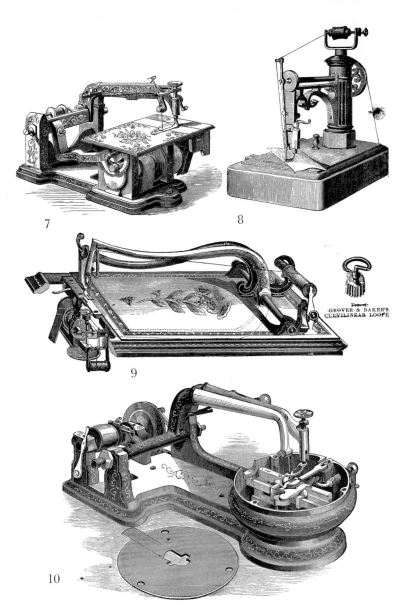

7   The Wanzer sewing machine.
8   Wright & Mann's sewing machine.
9   Grover & Baker's knotted stitch machine. Grover & Baker
     were Singer's chief competitors.
10 The Florence machine, with cut-out showing mechanism.

11   One of the first commercial machines produced by Willcox &
Gibbs Sewing Machine Co in 1857. James Gibbs was a pioneer
in the production of a less expensive sewing machine.

during his absence the sewing machine had made its own way
without him. He managed to redeem his sewing machine and
papers from the London pawnshop, but Thomas had obtained
all rights to it in Great Britain, and from his shop in Cheapside
he became the first purveyor of sewing machines in London.

By 1863 the price of the British sewing machine had dropped
drastically to between £5 and £20. £30 had been out of reach
of the seamstresses, but now they could obtain one to expedite
their trade. It had taken fourteen hours to make a shirt by hand;
a sewing machine could do one in an hour and sixteen minutes.
A lady's chemise took ten hours by hand, one hour by machine;

a dress, eight and a half hours by hand, fifty minutes by machine. Aprons could be made in less than ten minutes by machine.

The initial response of the British public was cautious, and at first it seemed that two or three shops would be enough to cater for the demand. The extravagance and luxury of the American showrooms were unknown in London. Yet within a year or two, shops were opening all over the country, and in Cheapside three adjacent shops were opened for the sale of sewing machines. The nearest approach to American advertising methods was the placing of a young lady machinist in the shop window, but this innovation was dropped when it was discovered that the only people who were watching her at work were young men with nothing better to do.

The first sewing machines were functional and bare, devoid of excessive ornamentation. This state of affairs did not last long, and with machines intended for the family market not only did the black japanning finish encourage ornamentation, but the actual shape of the machine was tormented into bizarre designs. The Singer family machine of 1858, dubbed the 'Grasshopper' by the salesmen, included an attractive design of flowers painted on the body, surrounded by a conventional border design. The arm was decorated with an inoffensive geometric pattern.

One of the first of the cheaper machines, Willcox & Gibbs' model of 1857, covered the entire frame with an attractive pattern of roses, the only non-moving part left undecorated being the plate on which the material rode. A pointer to the future was the Dolphin sewing machine of 1858, in which the frame was twisted into the form of a dolphin; the end of the tail winds back to the belly of the creature, forming a cross bar that is totally irrelevant to the machine. The true baroque sewing machine is seen in the 1858 Cherub sewing machine, manufactured by the same company that produced the Dolphin. The upright of the machine is formed by two gilt cherubs, one of which holds the spool, the other a cross bar connecting the upright with the needle unit. For some quaint reason, the needle mechanism is backed by a large gilt dragonfly.

The same firm evolved about the same time the Foliage sewing machine in which the machine is disguised as stylised leaves and branches, but they were not responsible for what is possibly the oddest sewing machine design of all. This is in the shape of a horse, with the front legs and head forming the needle unit, the back legs and hindquarters the upright, though another candidate for the oddest might well be the Sewing Shears machine which has the appearance of a pair of scissors clamped on a wooden stand. The spool projects above one of the finger-holes, and the other finger-hole is left free, for this machine could not only sew but cut as well, the scissors that form the design being actual scissors.

12   The Dolphin sewing machine of 1858. The reel of cotton gives some idea of the dimensions of this machine, which is made of solid brass and is five inches long.

An eccentricity of 1859, nominally a sewing machine, featured a mermaid holding the needle, two semi-draped figures decorating the upright, while the connection between the upright and the needle unit was furnished by two serpents. Normally the plate across which the material went was left undecorated, but in this case it was shaped like a heart.

Such adornment is only practised when the techniques have been conquered, and unquestionably the step from inception to perfection in the sewing machine was short. The only path to be pursued was in the provision of an increasing range of attachments. The addition of a power source was not so important in the sewing machine as it was to other domestic apparatus developed in the 1850s and 1860s, such as washing machines. Where mechanisation was necessary, as in factories, steam power was available almost from the start. It is interesting that an electric motor was added in 1871 to an 1865 machine, but this was too far ahead of its day to be a commercial proposition. Electric motors were first used to drive a fan.

Allegiance was divided between the treadle machine and the hand-operated machine, with perhaps the treadle model dominating, as it left both hands free for the manipulation of the material. It occurred to some makers that the treadle provided an excess of power, power that could be utilised for another purpose, and in 1882 an American patent was received for

a cover for a sewing machine provided with a musical instrument and means for transmitting motion from the shaft of the sewing machine to the operating parts of the musical instrument.

The musical instrument set in the cover of a sewing machine was of the nature of a small mechanical organ, using a paper roll to activate the mechanism. These organs, so-called organettes, were very popular in their own right. The same kind of principle was used in the more widely-known pianola.

In the early 1870s a fan attachment which was operated directly by the treadle was available for fitting to sewing machines at the cost of a dollar.

There is no question that the early sewing machines qualify as true domestic antiques. They are a salutary reminder of the sophistication of nineteenth-century techniques, and of the important contribution of the United States to the mechanisation of the home. Their workability and reliability depended on precision engineering and modern production methods. The assembly line came early to American industry, being initiated in the late 1790s by the gun-maker Whitney, and without mass-

13   *The Cherub sewing machine made by the same firm as the Dolphin, in the same year.*

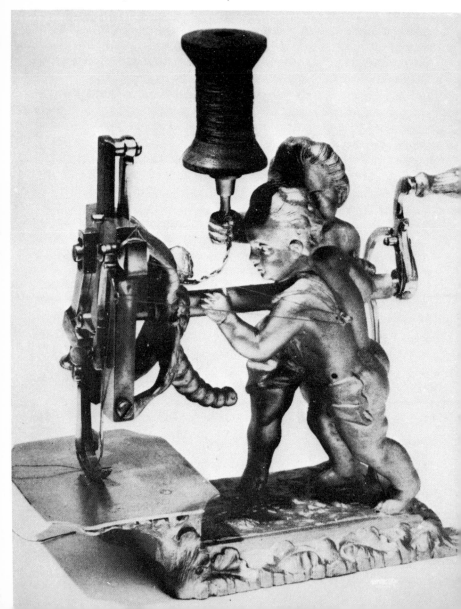

production there could be no question of the philosophy of inter-changeable parts. Without a supply of spares that were guaranteed to replace sewing machine parts worn or broken, the sales and service programme evolved by Singer would have been unthinkable, and this also applies to the mail-order business inaugurated about 1870. Sewing machines were prominent in the pages of the mail-order catalogues.

Quantity was the key to the success of the sewing machine. By 1860 there were seventy-four factories in the United States producing more than 111,000 sewing machines a year, with fourteen factories devoted to the production of cases and attachments.

The extravagance of the designs and the exuberant pursuit of technical variation were only possible in a country where there was a lot of money available to back and promote the products. This happened in Britain in the early 1850s with the promise and optimism of the Great Exhibition. American entries to the exhibition were praised for their simplicity and fitness for purpose, but a decade later the Americans were proving that they could be as extravagant in their designing as the British.

At one time sewing machines fashioned as dolphins, horses, or cupids would have been seen as worthy of contempt, and deserving of no better fate than the rubbish tip. However, we are no longer longer so puritanical in our attitudes towards the past, nor can we raise much indignation about the nineteenth-century division between Fine Art and Applied Art. The nineteenth-century sewing machine manufacturers thought that their products would be enhanced by dressing them up. They scorned concepts such as Art Manufacture, a term invented by Henry Cole in 1845; Cole maintained that 'an alliance between art and manufacturer would promote public taste'. The aversion to such objects as Dolphin sewing machines was put clearly by Herbert Read in *Art and Industry* in 1934.

> The necessity of ornament is psychological. There exists in man a certain feeling which has been called *horror vacui*, an incapacity to tolerate an empty space. This feeling is strongest in certain savage races, and in decadent periods of civilisation.

The mechanical object dressed up as an art form struck Herbert Read and his generation as lacking honesty, but surely a sewing machine in the guise of figures, serpents, and mermaids strikes one as ingenious and not offensive, if the sewing machine itself is a working proposition. One could just as well berate the makers of automata for producing a useless intricate piece of work that could be accomplished by more simple means.

The sewing machine, whether in its basic first stage or in its later more ornamental and involved forms, occupies an almost unique role in domestic and mechanical antiques, social evolution, and the history of technology. The claims of its devotees may strike a little odd on our ears—it is difficult to visualise clergymen treadling away as they write their sermons, or tiny tots making aprons on those occasions when they were allowed to use the machine.

Their familiarity makes it difficult to appraise them afresh, but if one looks at the modern streamlined sewing machine it is clear that this is exactly the same instrument Singer put on the market in the 1850s.

There are large quantities of sewing machines to be found in

14 This unusual machine of 1859 features half-naked female figures carrying the spool, a mermaid holding the needle, and a heart-shaped base.

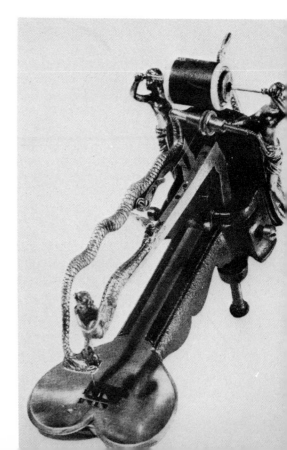

antique and junk shops, and the smaller they are the more expensive they are. Although functional, the small machines measuring five inches across give the impression of being miniatures, akin to the samples that furniture salesmen took about with them in the eighteenth century. Even these small machines are underpriced when one considers their decorative possibilities.

The exotics, sewing machines in the forms of dolphins or cupids, are quite rare in Britain. The Americans were going through the phase of exuberance that Britain experienced during the period of the Great Exhibition ten years before, and Britain was settling down to the fact that life was real and earnest. Nevertheless, the solidity of the sewing machine as seen in Britain does not detract from its collectability, and there is usually a degree of ornament.

Like other mechanical antiques, sewing machines can be deceptive—they may not be so old as one thinks, though the ornament is always a good clue to date. Art nouveau type decoration came in towards the end of the nineteenth century and lasted until the first world war, and the pretty-pretty or severely geometric ornament of the 1920s is easily distinguishable from Victorian or pre-first world war designs.

15 The powered sewing machine—a novel 1915 solution.

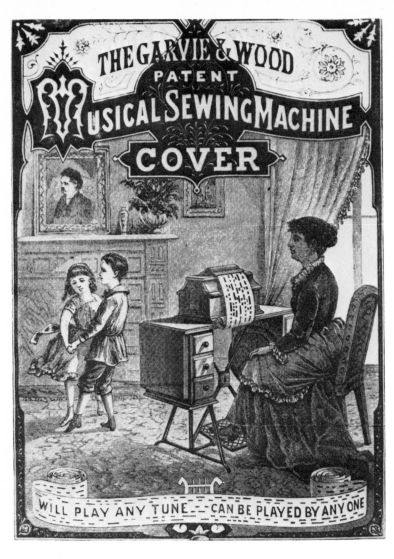

16   In 1882 a patent was received for 'a cover for a sewing machine
     provided with a musical instrument and means for transmit-
     ting motion from the shaft of the sewing machine to the
     operating parts of the musical instrument.' The principle of
     perforated music rolls was used in small mechanical organs of
     the period.

Fortunately a much better indication as to age lies in patent numbers. These are often hidden under a patina of dirt. Sewing machines that one rescues from junk shops are often in a sad state; a modest amount of rust should not deter one from buying, though it is a good idea to scratch away at any rust to see how far it has bitten in. Rust can be removed by branded products supplied by motor accessory shops. The mechanism should be tried out, in the event of one wanting to use the machine; unless it is exceptionally pretty it is inadvisable to buy a machine that has jammed solid. To many collectors' minds the machinery is as fascinating as the frame. There are enough machines about for a collector to be choosy. If the handle does not turn, and there does not seem any reason why it should not, it is a good idea to look at the hidden mechanism. There are usually two catches on the base of the machine; if these are slid across, the chassis of the sewing machine can be lifted on the hinges at the back. It is worth doing this anyway, for there is always the chance of finding interesting sewing accessories. Even the most ordinary of silver thimbles, sold for a shilling or two at the time, are worth upwards of £1.50.

CHAPTER 2

THE TYPEWRITER

ALTHOUGH THE first successful typewriter was not put on the market until 1873, numerous attempts had been made to create a viable alternative to handwriting. The need was psychological as much as economic, the need to speed up the methods of recording thought and speech. An early example of this desire was the invention of shorthand writing, which dates back at least to the fourth century BC, and was used by the Greeks. The medieval scribes formulated the practice of using abbreviations in manuscript writing, which filtered through into the early days of printing, while shorthand systems were revived in 1602 with the publication of John Willis's *Art of Stenographie* which ran to fourteen editions by 1647.

The adoption of printing in the mid-fifteenth century freed the western world from the tyranny of the pen, and it was only a matter of time before the intervention of an intermediary between the pen and the printing machine. In 1714 a patent was granted to Henry Mill, an English engineer, for

> an artificial machine or method for the impressing or transcribing of letters singly or progressively one after another, as in writing, whereby all writings whatsoever may be engrossed in paper or parchment so neat and exact as not to be distinguished from print.

No details of his invention have survived, nor is anything known of Mill other than that he had been 'engineer to the New River Water Company'.

17 A Type Recorder of the 1880s, a primitive British typing machine that did not attempt to compete with the sophisticated American products of the period.

A number of primitive devices was thought up, but there was little economic need for writing machines when ill-paid clerks wrote precise and legible hands, and duplicate records could be made with damp tissue and a copying-press. The basis of these early machines was the hand manipulation of one letter at a time, but a modest fillip was given to their evolution by the need to help the blind. Pingeron, a French inventor, made a machine in 1780 for the use of the blind, another similar machine called a 'writing frame' was evolved in 1784, but the first practical device was produced in 1808 in Italy by Pellegrino Turri. This printed in capitals 3 mm high, the paper being adjusted by hand; this was a one-off, constructed for a blind countess.

The first American patent (259) was granted to William Austin

Burt in 1829, and his 'typographer' or 'Burt's Family Letter Press' is the first typewriter-like object to survive. Although the original in the US Patent Office was destroyed in a fire in 1836, a replica was made from the original papers in 1892. Like Hunt, prominent in the evolution of the sewing machine, Burt was an all-round inventor; in 1836 he invented the solar compass, with which the true meridian could be ascertained by a single observation of the sun, an invaluable instrument for navigators. The typographer was a lettering device rather than a typewriter. The types were on the rim of a wheel, and by turning a dial pointer the required character was brought into position. The impression was made by depressing a lever. The types were inked by contact with inking plates, fed with a hand-roller.

More important was the invention by Xavier Progin, a Marseilles printer, of his 'Machine Kryptographique'. This 1833 machine embodied one of the two basic principles of the modern typewriter—the types were on separate bars arranged in a circle, and each type struck upon a common centre. Progin claimed that it would write 'almost as fast as a pen' and it was therefore by implication a failure. The structural disadvantage of Progin's machine was that the type struck *downwards* instead of upwards, and therefore the bars would not return to their station by gravity.

Nevertheless this was the most important achievement so far, and with the invention of Guiseppe Ravizza's 'Cembalo-Scrivano' we see something approximating to the first successful typewriters of the 1870s. Although Ravizza's first machine was produced in 1837 it was not exhibited until 1856. The 'Cembalo-Scrivano' was of the 'up-strike' variety, with type-bars hung around a circular type-basket, striking up to a common printing centre on the underside of paper carried in a flat rectangular frame. Although Ravizza had not hit upon the second basic principle of the successful typewriter—a cylinder holding the paper moving longitudinally to space the letters and rotating to change the lines—he does appear to have been the first to use a ribbon inking system (a silk ribbon impregnated with lamp black, and later with black lead paste and prussian blue dye). Ravizza was also one of the first to arrange his piano-like keyboard in order of usage rather than alphabetically. Had Ravizza's machine received the attention which it was due, Italy might well have been the centre of the typewriter producing business. It is a mystery why it had to wait nearly two decades before being given an airing.

Charles Thurber's 'Chirographer' of 1843 (US patent 3,228) was in one sense a backward step. The type was arranged round a rim, and depressed on to the paper by hand on the plunger principle. The 'Chirographer' is historically significant in that Thurber used a lengthways moving roller around which the paper was gripped, but in the early models the process was incredibly long-winded, for the roller had to be moved along by hand.

No machine of this period could in any way compete in speed with the pen, and although several machines were exhibited in

18   The Remington typewriter of 1878 remained almost unaltered for more than twenty years, except for the addition of a shift mechanism. Reliable and almost indestructable, many of these machines are still in working order.

the Great Exhibition of 1851, they caused little stir. There is no evidence to suggest that the increasing number of newspapers and periodicals—in 1840 there were 493 newspapers in the United Kingdom—accelerated the demand for something faster and more legible than the pen. The widespread use of Pitman's method of shorthand, introduced in 1837 under the title steno-graphic sound-hand, and taken up by reporters with avidity, was not associated with an equally rapid means of transcription. Transcription was bypassed by handing the shorthand notes directly to newspaper compositors; in 1845 an hour and a quarter long speech by Richard Cobden in Bath was reported in short-hand, and the notes given to the *Bath Journal*, the compositors for which set up the text from these notes with complete accuracy.

There was one area in which a typewriter would have been extremely useful. In 1851 Sir Charles Wheatstone produced the first of a series of machines for the rapid printing of telegrams, but his machines were out of the main stream of design. The British, French and Italian typewriters were produced by dilet-tantes, many of whom felt that they had a moral duty to provide an instrument for the blind. Financial and commercial incentives were lacking, and it is not surprising that whereas European experimentation tailed off when it seemed that the public was not interested, the Americans pursued the chimera with diligence and energy. Although there were many problems to be overcome, the example of the sewing machine had shown that if a truly marketable machine could be produced sales possibilities were enormous.

The eccentricities fell by the wayside. J. B. Fairbanks's 'Phonetic Writer and Calico Printer' was one of these, though it did introduce the continuous roll paper feed. Oliver T. Eddy's 1850 machine had seventy-eight keys disposed in four rows after the fashion of an organ manual. The types were formed on the ends of vertical square plungers massed together as a rectangle of bars, moved by a selector mechanism. The selected bar was thrust down upon the paper by a vertical piston rod. Complicated and passably workable, it was as big and intricate as the later linotype machine.

One typewriter that did point the way to the future was that patented by Samuel W. Francis of New York City in 1857.

Francis was a wealthy doctor, and he made no serious attempt to promote his machine. This had a piano-like action; hammers were hung in a circle, and struck from beneath at a common point, over which a small circular platen was suspended. It was extremely bulky, and the parts were delicate. But it wrote cleanly and rapidly. It had two disadvantages—the parts were too fragile to stand up to continuous use, and it could not have been manufactured to sell at a reasonable price. Mass-production was the secret of success.

By 1860 all the elements of the successful typewriter were there, in particular the two basic principles: a circular arrangement of type bars striking at a common centre, as used by Progin in 1833, and a cylindrical paper carriage, as used by Thurber in 1843. Only the man to correlate all the information was wanting.

This was Christopher Sholes, an American printer living in Milwaukee, who had in 1860 designed and built a machine for addressing newspapers to customers using addresses already set up in type. In 1864 he patented a numbering machine, almost identical with the numbering hand stamp of today, with types on the periphery of rotating discs. In 1867 he read an article on typewriters in the *Scientific American*, and from the brief description of a model being exhibited in London he began to evolve his own, at that stage being unaware of previous developments in this field in America. He first of all used parts of a telegraph sender to construct a device which would tap out just one letter, and gradually built on this until he had a workable machine. This typewriter was as primitive as earlier machines, but it struck the imagination of a reporter, Charles Weller, who ordered the first model, and used it for two years until Sholes came up with something better. Despite the manifest faults of this typewriter, Weller preferred it to the pen. Although the impregnated ribbon had been evolved in Italy thirty years previously, Weller had to make his own, soaking strips of silk or satin in ink. A spring had already been used for moving the carriage, but Weller had to make do with a weight.

Sholes was fortunate that at this stage he found a backer, a former colleague, James Densmore, who paid Sholes and his helpers $200 each and agreed to finance the manufacture of the machine before he had even seen it. The crude appearance of the typewriter shocked him, and he demanded improvements, but these did not satisfy him and in 1868 he withdrew from the undertaking. Only after Sholes had examined other machines and learned from them did he return to the scene.

Sholes decided to have the cylinder of his new model rotating instead of sliding lengthways to space the letters. The machine would write around the cylinder as it revolved. It worked, and he wrote to Densmore: 'Its simplicity cannot be equalled, it being more simple if possible than a piano and so less liable to get out of order.' At present he could do twenty words a minute, but he admitted that he was old and clumsy, and sixty words a minute could not be out of the way for a younger hand. He drew a shrewd analogy with the sewing machine.

> Sewing machines, which are now made complete at a cost of somewhere from ten to fifteen dollars, are better made and have as much machinery about them as some of the first ones made, which cost several hundred dollars each. I anticipate a similar experience in this case.

Sholes was typical of a breed of American that rose in the 1850s and 1860s, capable and far-sighted, and in tune with the American dream. In particular, he had enthusiasm. Densmore had the contacts; he was also more cautious, considering that the new Sholes model could be bettered, and he gave the model out to various reporters instructing them to work it to the limit. A stenographer, Henry Roby, pounded it until it would no longer 'co-operate in its several parts'.

Such consumer research was ahead of its day, and as defects and design errors came to light, Sholes re-invented and altered, until in 1870 his typewriter was taken to New York. Just as the telegraph companies in England were interested in Wheatstone's typewriter, so did the American telegraph companies view Sholes's machine with an eye to printing telegrams rapidly and clearly. One of the experts recruited by the Automatic Telegraph Company was Thomas Edison, then involved in trying to record Morse code by chemical means. Edison did not think highly of Sholes's machine and set to work to make one of his own, which was, however, slower and more complex, though it provided the ancestor of the ticker-tape machine.

Discussions with the telegraph company fell through, but Sholes was still sanguine and continued to develop his machine,

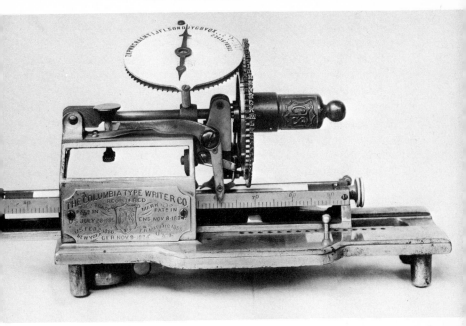

*19 The extremely decorative Columbia typewriter of 1886, which
although later than the Remington, reverts to a more primitive mechanism.*

evolving buttons instead of a piano-style keyboard, and inventing
the concept of the space bar. The Automatic Telegraph Company came through with an order for a small number of
machines, and although matters did not yet warrant mass-
production, twenty-five machines were made in 1871. Densmore
was running short of money. He had spent $10,000 on develop-
ment, and felt that the whole project had been dogged with mis-
management and ignorance. Forty separate and distinct machines
had been made, no two alike. But he wrote, 'I believe in the
invention from the top-most corner of my hat to the bottom-most
head of the nails of my boot-heels.'

During the winter of 1871–2 Sholes decided that the roller
should run horizontally rather than rotate, and in his latest model
he incorporated a treadle which would return the carriage and
turn up the platen a notch to start a new line. A bell was added
to give warning of the end of a line. New York businessmen

were encouraging Densmore, promising a greater sale than the sewing machine ever had.

In 1872 Densmore opened a factory in Milwaukee, and the firm received valuable publicity by a front-page article in the *Scientific American*. The factory was old-fashioned in that it made no use of production flow, and each machine was treated as a separate task. There was no interchangeability of parts, and each component had to be filed and bent for its specific role in one typewriter. The now obsolete 'axle' models (roller rotation, not horizontal movement) were cannibalised to make new machines, and no two machines were alike. Except in one respect. Sholes rationalised the keyboard; the pattern was QWERTYUIOP exactly as it is today.

The machine built in Milwaukee was basically the modern

20 *The Crandall typewriter of 1881, superbly designed and technologically ahead of its time, using a type head instead of bars, thus anticipating the IBM 'golf-ball'.*

typewriter. There were only two improvements to make—the provision of a shift key to have lower case as well as capital letters, and a front-striking movement so that the operator could see what he or she was typing. The shift-key was introduced in 1878, and the front-striking movement in 1880.

All the machines produced in 1872 were sold, mainly to telegraphers and reporters and a few professional men. Densmore was disappointed that businessmen shunned the typewriter. Also the *ad hoc* system of manufacture was making the machine almost prohibitively expensive; they were costing more to make than the selling price.

In December 1872 George Washington Yost visited Chicago, and was invited by Densmore to the Milwaukee plant. Yost was a tycoon in the shipping business, who owed Densmore money. Milwaukee was no place, he maintained, for making typewriters, but out of a sense of obligation he put Densmore on to the place which was, Ilion, New York, and the people, E. Remington & Sons, the gunmakers.

The gunmakers of America were pioneers in the techniques of mass-production. Samuel Colt (1814–62) was determined to apply the principles of interchangeable manufacture at all stages of the production of his revolver, and with the help of a brilliant engineer, Elisha Root, the production of standardised components was brought to fruition at the Colt Armoury, Hartford, Connecticut, between 1849 and 1854. Root designed and built numerous semi-automatic machine-tools, and made large quantities of gauges to ensure accuracy. The components produced could be accurate to a millionth of an inch.

Remingtons had already diversified their output, were producing sewing machines at the rate of fifty a day, and were preparing the production lines for a steam plough. They agreed to remodel the typewriter to suit mass-production, to manufacture at least a thousand, plus twenty-four thousand more at their discretion. Densmore and Yost agreed to pay Remingtons a fixed sum for each machine, plus $10,000 as an advance payment. Within a few months a pattern machine was produced, but the whole of American industry was hit by a slump. Densmore set up his sales headquarters in a dingy street in Manhattan; his letter-heading announced 'The Type-Writer. A machine to write with types and supersede the pen for common writing. Price $125.' There

45

were no longer problems with the ribbon; this could be used for months, and when it needed re-inking it was sent back to the sales point.

In April 1874 Densmore received his first Remington typewriter. The machines manufactured at his Milwaukee factory gave the impression of being hand-made. Mass-production had made of the typewriter a much more handsome artifact, wood was replaced by black enamelled metal, with gilt script identifying it. The head mechanic of Remingtons had been in sewing machines, and the typewriter used the same treadle with its grape-vine pattern.

In December 1874 Mark Twain bought one, but sales did not go well and between July and December 1874 only four hundred machines were sold. There was considerable friction between Densmore, Yost, Sholes, and earlier helpers, and in November

21 *The Oliver typewriter of 1894, using the sideways motion of the key bars. Like the Remington, the Oliver was well made and is still to be seen in working order.*

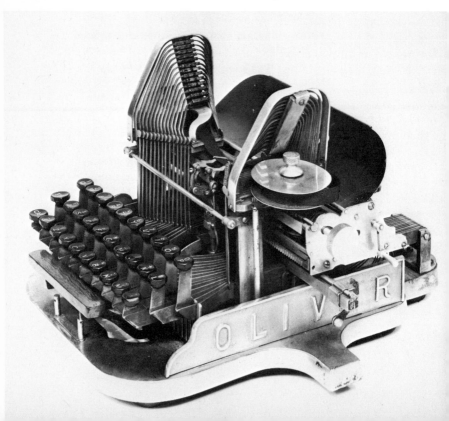

1875 the newly-formed Type-Writer Company, comprising all those involved in the Sholes machine, granted Remingtons the exclusive right to make and sell their machines. Yet even the sales force of Remingtons failed to make an impact on the uninterested public, and a specially built model rich in mother-of-pearl did not get noticed at the Philadelphia Exposition of 1876.

Sholes began working on a portable machine, while Yost designed a machine of his own, the 'Caligraph', on sale in 1881. It had a number of interesting features. The polygonal platen, exposing a flat rather than a curved surface to the type face, was a detail, but to print both capitals and lower case Yost used a keyboard with seventy-two characters. For its day it was cheap—$60 for capitals only, $80 for the full keyboard.

Remingtons drastically remodelled their sales operation to cope with this threat, and the war between the typewriter companies was joined in earnest, accompanied by patent evasions and lawsuits. It was won by Remingtons; in 1880 they sold 610 machines, in 1881 1,170, in 1882 2,272, in 1883 3,376. Other competitors entered the scene. The Hammond, one of the more interesting, used a type-wheel rather than individual types on a rod. All the types were on a curved plate on the rim of a horizontal wheel, the particular type would be brought into position, and a hammer would strike the paper and the ribbon against the stationary type face. Slower than the now orthodox machines, Hammond's 1880 machine had one advantage—the curved plate exhibiting the type could be replaced by other plates with different founts or characters.

The Crandall machine, patented in 1881, is interesting in that it anticipates the modern 'golf-ball' method of IBM typewriters. The types were on a polygonal sleeve that fitted over an arm. At the touch of a key the sleeve twirled about and moved up or down, then struck the paper. The Crandall was exceptionally advanced for its day, but lacking the astonishing speeds possible only on electric typewriters it remains a technological curio. Its price was only $60, but cut-price manufacturers were now making instruments for as little as $12, though these were operated by sliding or rotating a character into place and then pressing down, a method then forty years out of date.

Despite their supremacy in the field, Remingtons were in deep financial trouble; the years of peace that had followed the Civil

War had not helped their arms trade, and the supremacy of Singer had limited their sewing machine market. E. Remington & Sons were taken over by another company, and although re-named the Remington Standard Typewriter Company, the Remington family was no longer concerned with typewriters.

Up to 1886 an estimated 50,000 typewriters had been made. The demand was much greater than the supply, and Remington retained the lion's share of the market although Underwood pro-duced the first successful front-striking machine which enabled the typist to see the work in progress, in 1893 (the designer was associated with the first front-striking typewriter, Yost's 'Caligraph' of 1880). In 1889 the Bar-Lock typewriter was paten-ted, and in the same year a machine using differential spacing was evolved by the conjuror J. N. Maskelyne. This was expensive to produce, and was too far ahead of its time for general adop-tion.

A curio that enjoyed surprising success was the Yost Type-writer No 1 of 1889. The automatic ribbon return had not yet been perfected, and imperfect ribbon movement—as well as the need to circumvent patent rights—encouraged Yost to evolve a typewriter using an ink pad. The Americans did not take to the Yost, but large numbers were imported into Britain, and they can still be seen.

An inking pad was also used for the Williams typewriter of 1892, the smallest practical machine then on the market, Sholes's portable having come to nought. This had what is known for obvious visual reasons as a 'Grasshopper' movement. One of the most noteworthy machines from the aesthetic viewpoint was the Oliver (1894), the only successful contender in the Down-Strike-From-Side class. The type bars consisted of two banks of hoop-shaped loops to the right and left. Everything is visible, and the machine gives the impression of an old fashioned cinema organ on a small scale.

The problem of the automatic ribbon reverse was solved by Remingtons in 1896 with their Remington No 7. This typewriter also incorporated a margin release key, and two years later Remington had a decimal tabulator on their commercial models.

The number of new machines produced, which did little the established products did not do, is explained by the running-out of earlier patents, and obscure manufacturers wanting a share

22 *A 1901 advertisement for the Smith Premier type-writer.*

of the pie. The English makers did little to break into the vast typewriter market. The English Typewriter Co of 1890 which was taken over by North's Typewriter Manufacturing Co in 1895, the Waverley, the Moya (1903), the Salter (1913), none of these British based machines more than scratched at the profits of the great American companies.

Although the portable typewriter dated back to 1893—the Blick—the first commercial success was the National which appeared in 1916. The major makers were always responsive to a trend, and the Remington Portable appeared four years later. 'Noiseless' typewriters (a thump instead of a click) were made from 1910 onwards, and these were partly a British innovation.

The Germans were very strong in the manufacture of typewriters for the European market, though they did not enter the field until the American industry was well-entrenched. In 1896-8, the Lambert typewriter was made by the German Gramophone Co of Hanover, and in 1898 Wernicke, Edelmann & Co of Berlin produced the Edelmann typewriter. This machine was built on the type-wheel principle, and is interesting in that it had three shift keys, giving capitals, lower case, and numbers. The American Keystone machine was manufactured in Germany under the trade name Grundstein; aimed at the popular end of the market, the Grundstein comprised only 156 parts. The Ideal typewriter of 1897 was patented by a New York company, but the actual manufacture was undertaken by Seidel & Naumann of Dresden. It was fitted with a useful innovation, the backspacer.

The social consequences of the typewriter were far greater than those of the sewing machine. Commercial schools and colleges sprang up to cater for the army of 'typewriters' (not the machine but the typists) anxious to enter the commercial scene. Male typists held all the speed championships but gradually women took over the role of typist. A shop assistant in 1886 earned six dollars a week; a typist fifteen. Via the typewriter women were breaking into men's preserves. The spirit of competition was abroad, and with the new art of dictation came a reappraisal of shorthand systems, especially Pitman's. Competition was encouraged by the typewriter manufacturers, and in 1889 the London agents of Remingtons ran a speed contest at their offices in Gracechurch Street. The first prize was a Remington

THE "MERRITT."
THE PEOPLE'S TYPE-WRITER.

This .is exact copy of The "MERRITT'S" work.
It is equal to that of any High Priced Typewriter. Relieves fatigue from steady use of
pen. Improves spelling and punctuation. Interests and instructs children. The entire corres-
pondence of a business house can be done with
it. Learned in a half hour from directions.
Prints capitals, small letters, figures and
characters, 78 in all. Price £3.3.0. complete.
Address- RICHARDS, TERRY & Co. Limited,
46, Holborn Viaduct, London, E.C.

23   *An end of the century attempt to produce a cheap alternative
to the more expensive standard typewriters.*

No 2 typewriter, nickel-plated, value £29, and in addition £20
was offered to anyone who beat the world's speed record of
98.7 words per minute set up in Toronto by Miss Orr. Four out
of the seven prize winners were men, and although Mr Oliver
Wren did not top Miss Orr's record, he averaged 79 words a
minute (411 words written, 16 mistakes, total words exclusive of
errors = 395).

Although it took longer than in America for the female typist
to be accepted in Britain, the social consequences were even more
dramatic. As secretaries girls began to have a status that they
had never had before, and well-bred girls who had to go to work
found that there was a vocation for them in the office, whereas
previously they had been forced to work as governesses or teach-
ers, patronised and underpaid. Emancipated young women of

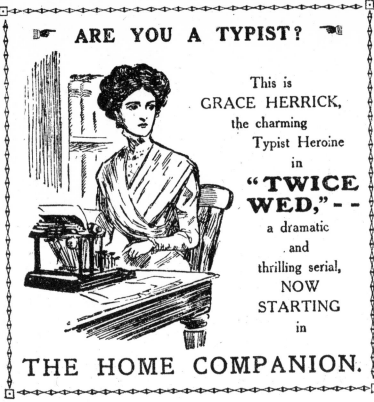

ARE YOU A TYPIST?

This is
GRACE HERRICK,
the charming
Typist Heroine
in

"TWICE WED," - -

a dramatic
and
thrilling serial,
NOW
STARTING
in

THE HOME COMPANION.

24 *The arrival of the typewriter brought in a new kind of girl—
the Typist Heroine.*

the leisured classes no longer had to remain in mute obedience
to their families; it was considered perfectly respectable for a
typist to go to the office without a chaperon. The influx of typists
and lady clerks into the City and the business districts of the large
towns created not only a new kind of commercial environment,
but a new kind of eating-place. Tea-rooms sprang up to cater
for this new breed, predominant amongst them the Aerated
Bread Company (the ABC) Restaurants. Without the typist the
cafeteria would have taken a longer time in coming. The presence
of women in offices was also a civilising influence.

   As collectable objects, typewriters have a certain cachet. There
were never machines as *outré* as the Dolphin or Cherub sewing

machines; the latter were designed to appeal to women. The typewriter was a commercial object, selected and paid for by men with an eye not to its decorative possibilities but to its utility. Oddities of design there were, but these were in response to technical needs, and although there was a certain amount of applied ornament on the bodywork, black lacquer being irresistible, only the treadles of the earlier machines (used for returning and turning the platen, and soon discarded) were amenable to true ornamentation.

Unquestionably nineteenth-century typewriters are decorative and fascinating for all with the slightest interest in technology. But are they workable? Are they useless objects, talking points, discards of a bygone age? Decidedly not. The ethics of nineteenth-century manufacture disdained the concept of built-in obsolescence, and mass-production included reliability in its curriculum. The typewriter of the age is no less likely to work than precision jobs such as the Remington pistol or the Colt revolver. Components were engineered on machine-tools that were no less accurate than those used today, and they were made to last.

Some of the smaller companies that arose when it was seen that the major companies could not produce all the machines demanded of them may be more suspect, and the $12 typewriters are hardly more than toys.

The biggest problem encountered if one wants to use old typewriters is the difficulty of getting ribbons. The Remingtons of the early years of this century used a ribbon approximately $1\frac{1}{2}$in wide. Until 1955 the author used an 1896 Remington No 7, and had no difficulty in getting ribbons.

Old typewriters are sold for function rather than display, and are to be found in office equipment shops quite as frequently as in antique or junk shops. From the technological point of view they are perhaps more interesting than sewing machines, for there was basically one sewing machine principle whereas the typewriter encouraged experimentation. The great thing about typewriters is that all the working parts are on view, and a fault can be traced through. The most likely fault on an old typewriter is a broken spring, but do-it-yourself enthusiasts will have little difficulty in adapting a modern spring for a machine.

Dirty and woe-begone as old typewriters may appear when

found, they spruce up remarkably well. The mechanism is very sturdy, and a good going over with a fairly thin machine oil can work wonders. The paintwork is almost always of high quality. It is advisable to have a good look at the typewriter frame, which was cast, and is therefore subject to cracking if treated badly.

CHAPTER 3

THE TELEGRAPH

IT MIGHT be supposed that the electric telegraph is no field for
a collector to venture into, and that the equipment relating to
it is either mundane or completely incomprehensible, and far too
esoteric to interest other than electrical engineers. This is very
far from the truth. The Victorians were excited and exhilarated
by anything that savoured of electricity, and the gadgets used
were invested with enormous significance. Fitness for purpose,
happily, was not an ethos that appealed to many, save the great
structural engineers such as Brunel, and the items surrounding
the new telegraph phenomenon were designed to have not only
engineering but aesthetic appeal.

During the first half of the nineteenth century the communication scene was turned topsy-turvy. It was an accumulation of
factors rather than one dramatic breakthrough, a transformation
effected by steam, electricity, increased facility in the handling
of metals, not to mention accidental by-products of the Industrial
Revolution and the bringing into action of theories that scientists
had held for centuries.

The telegraph was a vital ingredient in these improved communications. It was born in the Napoleonic Wars. During 1793-4
the French Revolution was at its height, France was assailed on
all sides by hostile nations, Marseilles and Lyons were rebelling,
and a British fleet held Toulon. France had one major card left
—the cohesion of the French Army, and it was to maintain
communication between the various sections of the French Army
that Claude Chappe was instructed to establish a line of tele-

graph stations between Paris and Lille, a distance of 144 miles. There was only one viable form of telegraph—visual, and so Chappe (1763-1805) built fifteen stations, each within sight of the next. This semaphore telegraph used a wooden beam pivoted at the centre, and at each end of the beam were two movable arms. The H-shape resembled a television aerial. A great number of configurations was possible using this method, but the semaphore system would not have been practicable without the telescope, an instrument that had been in existence since 1608.

Primitive as Chappe's system was, it did the job successfully, and within a few years lines were established between Paris and Strasburg and Paris and Brest. Where the terrain was undulating there could be great practical difficulties in siting the stations, and between Paris and Strasburg there were fifty stations. Nevertheless, the system was constantly being added to, and by the time it was obsolete in the early 1850s the network covered more than 3000 miles and included 556 stations.

The British were not slow in copying the French lead, but selected a framework with six movable shutters for its stations, fifteen of which were demanded for the first route, between London and Deal. This was in the nature of a pilot scheme, but when it was found that a signal could be sent and acknowledged in two minutes it was considered a good idea to have a telegraph system between London and Portsmouth. Prominences between these places still go by the name of Telegraph Hill, clear evidence of their usage. London was also linked with Yarmouth and Plymouth, but it took about fifteen years for the British to adopt the French system, for although the shutter system could frame sixty-three different signals it was cumbersome and confusing.

War had made the semaphore telegraph worth-while. In Britain, the Admiralty had been the main users. But the cost of maintenance of the many stations and the waste of manpower caused the stations to be run down during the peace, and the re-opening of some of the routes during Napoleon's revival in 1815 was only temporary.

Electricity in the early nineteenth century was something of a mystery. Natural electrical phenomena were sporadic, disconnected, and frequently trivial; the attraction of amber to straw was a scientific curiosity. Lightning was impressive but perplexing, lightning 'which doth cease to be ere one can say "It lightens" ' (*Romeo and Juliet* Act II Sc 2). Electricity was an amusing extension of its sister science, magnetism; magnetism could be used (as in the compass) or misused (as by Mesmer who gave the world a new word, mesmerism, when he asserted that magnetism was therapeutic). But all that electricity could be used for was trifling experiments.

The century had opened with the discovery of the electric current, but few saw the possibilities and potential. In 1820 Oersted discovered the effect of electricity on the magnet, giving them a common basis of theory. But it was still the stuff of parlour magicians. Brunel declared that electricity was only a toy, echoing eighteenth-century views. The eighteenth century had seen electricity used for tricks; the discovery of the Leyden jar in 1745 meant that electric charges could be stored, but the only use this was put to was the giving of mild electric shocks to sensation seekers, culminating in the grand coup of Abbé Nollet (1700–70) who put 200 monks in a circle, connected them with iron wire, and passed a shock, a mile long, through them.

That electricity could be more than a means of titillating the curious was demonstrated by Francis Ronalds in 1816. Ronalds, born in 1788, had been dabbling in electricity for four years, and thought that somehow electricity could be used in place of the laborious telegraph system using semaphore. Ronalds lived in a house in Hammersmith, London, the garden of which was 600ft long. He erected a pair of large wooden frames, from the bars of which he suspended eight miles of wire. One end of the wire was connected with a frictional machine which charged the line with electricity; to the other he connected an indicator, two pith balls which moved when the line was charged. At the sending station there was a dial marked with the letters of the alphabet, and this rotated behind a plate which had an aperture enabling one letter to be seen at a time. The receiver had an identical dial, synchronised with the first dial by clockwork. The line was charged continuously; when the desired letter appeared, the line was discharged. The power was removed, and at the other end of the eight-mile length of wire the two balls of pith would indicate to the receiver that the power was withdrawn. He would note the letter appearing in the aperture. The principle of the electric telegraph was born—the communication of information by the sender cutting off the electric current.

26 *The Cooke and Wheatstone five-needle telegraph of 1837.*

Not content with this, Ronalds proceeded to try to see if the experiment would succeed if the line was placed underground. Using copper wire insulated with glass tube, he placed 525ft of this wire in a trench 4ft deep. He was sufficiently pleased by the success of his two experiments that he approached the Admiralty, but the Admiralty, no respecter of innovation, declared that their present system of telegraphy, with men in cabins scattered over the south coast equipped with telescopes, was good enough for them. Disconcerted, Ronalds turned to other fields, and in 1845 he invented a system of automatic photographic registration for meteorological instruments.

If Ronalds does not belong to the electrical pantheon, he is certainly close to it. But although he had found the principle he had not found the method. Hans Christian Oersted, professor of physics in the university of Copenhagen, found that a magnetic needle was deflected when current flowed through a wire above it. To outsiders an innocuous happening, it caused great interest to all those connected with electricity, and in due course an electric telegraph system was evolved using electromagnets.

All this was happening in Europe, but in 1836 William Cooke, realising the possibilities of the electric telegraph, constructed several forms of telegraph. The railways were very interested; a telegraph system was almost essential for their efficient working. But Cooke ran into difficulties in the working of the electromagnets, and asked for help from Charles Wheatstone, professor of natural philosophy at King's College, London, who had himself been conducting similar experiments, and who was so impressed by the calibre of Cooke's work that he proposed a partnership. The first patent granted to Cooke and Wheatstone was dated 1837.

The first successful electric telegraph in America came out in the same year, the brainchild of Samuel Morse, a painter and amateur physicist, but America was going through a mild slump and Morse could not get Congress to advance him the $30,000 he needed to promote the telegraph. The possibilities of it were not seen; the railway age had not yet arrived in America. Morse had to wait until 1843 before his telegraph system was accepted, and the first line was laid between Washington and Baltimore, a distance of nearly forty miles, in 1844.

Three things were necessary for an efficient telegraph—a

27 *The Cooke and Wheatstone two-needle telegraph.*

means of making a current, a means of indicating precisely when the current was discharged, and a code. The electromagnet was the means, but there were three ways of indicating the discharge of current—a dial and an aperture in a plate, as in Ronalds's experiments, a bubble of hydrogen, or a needle, which would point to the letters. Wheatstone favoured the needle method, and the first telegraph used five needles. This was for a mile-long line between Euston and Camden Town. This object is one of the first artifacts *designed* for its purpose, rather than simply made. Earlier experimental apparatus had been put together on an *ad hoc* basis, and although all early electrical apparatus is interesting and should be collected, the Wheatstone five-needle telegraph of 1837 is the first true antique of the telegraphic age. In appearance it resembles an early Victorian toy, and the wooden carving about the lozenge-shaped dial is characteristic of the period.

A line between Paddington and West Drayton, thirteen miles, was completed in 1838, and proved so successful that it was extended to Slough in 1842. This was a two-needle telegraph. The five-needle telegraph had proved cumbersome; instead of having the needle point to a specific letter on a dial, would it not be better to have a code? Especially as an exceptionally efficient one had been evolved in America by Morse, the familiar dot-dash method. However, information by intervals rather than the common-sense way of having a needle pick out a letter was too novel, and the ABC telegraph, as the five-needle telegraph of Wheatstone was known, remained popular for many years, though immeasurably slower than telegraphs using codes. The operators did not necessarily use Morse code; many evolved their own. In the early days the telegraph was an accessory of the railway, and consequently information could be abbreviated and codified to suit the exigencies of the moment.

The two-needle telegraph of 1842 is an object that would be attractive in any setting, with fretted carving, and Grecian pillars as side-supports (a device also used in steam-engines). As more lines were built so did the equipment accumulate up and down the country. Norwich was connected with Yarmouth, London with Southampton, Tonbridge with Maidstone. Two-needle telegraphs were followed by single-needle telegraphs.

So far the public had not taken much interest in the telegraph;

28 *Another variation of the Wheatstone alphabetical telegraph.*

if they knew about it it was as just another complicated gadget for the use of the railway companies. They marvelled at the apparatus, an industrial antique of the future, describing it as 'a mixture of the beer-engine and the eight day clock' or 'the differential calculus framed in a gooseberry tart'—the Victorian man in the street was no respecter of scientific hardware. An event in 1845 struck the public imagination. A woman was murdered at Slough and her murderer was seen to board the train to Paddington. A full description of the man was telegraphed to Paddington, and when he arrived he was recognised, arrested, and, eventually, hanged. Nothing was more calculated to excite the masses, though it would be cynical to speculate that it was all cunning public relations with a human sacrifice thrown in for good measure.

It was clear that the telegraph was more than an accessory to the railway industry. The public were courted, invited to send telegrams for sixpence each. They were puzzled by it, intimidated, sometimes frightened. A man writing a telegram thought that

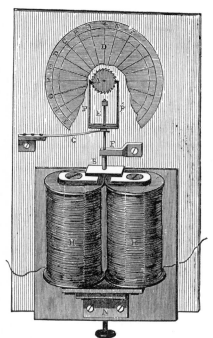

29   *Diagram of Wheatstone alphabetical telegraph.*

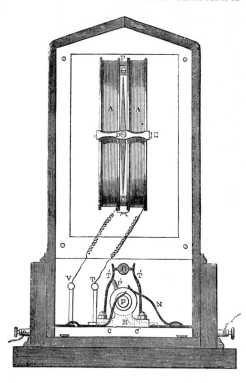

30 Diagram of single-needle telegraph.

in some magical manner the writing would be transported through the wires in a fluid state. A woman asked for two dozen stamps to be transmitted. A woman who lived by telegraph wires complained not of the voices she believed were being transmitted, but of what they said.

In 1846 the Electric Telegraph Company was formed; by 1852 there were about 4000 miles of telegraph-line in Britain. New and more efficient apparatus was forthcoming, mainly invented by Wheatstone who did his best to thrust Cooke into the background. In other countries different methods were being employed. Samuel Morse invented a receiver that consisted of a pen held in contact with a moving strip of paper; when the receiver was energised the pen was moved by an electromagnet, and a notch appeared in the line drawn by the pen. Although Morse's instrument had to be redesigned by more practical men, it proved successful, and in 1844 a telegraph was installed between Washington and Baltimore, a distance of forty miles, and

65

31 'A mixture of the beer-engine and the eight day clock', a telegraph as used in the House of Commons in 1846.

companies proliferated, until merging into Western Union in 1856.

In Britain it was not the railways which underwrote the success of the telegraph system, nor was it the public, who used it solely for news of distress and death, but business. Instant communication was profit. It is therefore not surprising that the government, pushed by business, were keen on getting telegraphic communication not only between London and Birmingham but across the oceans. Twenty years after the first faltering start between Euston and Camden Town, the attempt was made to bridge the Atlantic by telegraph, using Brunel's *Great Eastern* to transport and lay the enormous quantity of submarine cable. A cable had linked France and England in 1851, and Scotland and Ireland in 1853. The difficulties in laying the submarine cable across the Atlantic were immense, but instant communication between Britain and the United States was essential to the economic prosperity of both countries (though instant communication was perhaps a misnomer—the maximum rate across the Atlantic was two and half words per minute).

Although the equipment being used in Britain still included the old ABC system of Wheatstone and Cooke, there were other methods in the offing. One of the disadvantages of the orthodox telegraphic system was the uneven quality of the operators, especially those who were only just beginning to come to terms with the Morse code. Puzzled recipients of telegrams were told that they would be met at the station by a

wig (·– – ·· – –·) —as opposed to a
cab (–·–· ·– –···)

or to expect a hog to dinner (actual examples cited by disenchanted Victorians).

The difficulty would be eliminated by getting rid of the human element. The first step in automatic telegraphy was invented in 1838. This ingenious instrument used electrified needles which were brought into contact with rolls of chemically treated paper which were fed into a transmitting mechanism and came only a year after the introduction of the five-needle telegraph system, with the equipment looking as though it had inadvertently escaped from the nearest Victorian home. Surely these are grounds for thinking that to the public of the time, automatic telegraphy savoured of science-fiction. The experts thought so, and the inventor was forced by financial difficulties to abandon his work and emigrate to Australia, though a quarter of a century later the idea was resurrected. In 1864 a patent was granted for an automatic transmitting apparatus, in which perforated paper was fed into a transmitting mechanism, and the messages were recorded at the other end of the line by a chemical recorder. Experimental transmissions were made, and it was discovered that up to 400 messages an hour could be sent. Automation of this calibre was bewildering to the Victorian mind, which could not cope with the continual products of the technological revo-

32   A more compact telegraph apparatus.

lution. The breakthrough in the practical use of electricity had triggered off a chain of events. This machine, little known except to historians of industry, was a pointer to the future, and although the 1864 machine was abandoned, the use of perforated paper for operating automatic machines was taken up by Wheatstone in 1866, found capable of handling up to 80 words a minute over a circuit 280 miles long, and by 1880, 170 of these machines, the direct ancestors of ticker-tape machines and all the data systems using the punched card, were in operation.

It will be evident that telegraphy in the nineteenth century was a big operation, with 8000 miles of cable and wire by 1854, and a number of companies vying for custom. There was no rationalisation of equipment, and a good deal of the hardware is still about, uncomprehended and incomprehensible, but superbly made. The two basic pieces of equipment for telegraphy were the transmitter and the receiver; as the century went on, the earlier five-needle and two-needle telegraphs were dropped in

33   An ambitious attempt to combine the telegraph with a printing mechanism.

favour of the Morse system, with the familiar Morse tapper, though there were many variations in techniques until the Post Office took over the whole of the telegraph system in 1870.

One of the problems for the engineers was whether to have overhead lines or underground cables. The chief disadvantages of overhead lines were the difficulties in getting permission to cross property, and the loss of power caused by inefficient insulation. The perplexities of medieval theologists in deciding how many angels could dance on a pin head were mirrored by the laws of freehold. How far did freehold extend *upwards*? To the sky? These were field days for lawyers concerned with property—though most people were content with a nominal shilling a year when the overhead lines crossed their houses or land. The underground cables also shared the problem of inefficient insulation. Gutta-percha (rubber) was used extensively. This was a material only recently introduced into industry, but it was used without sufficient trials, and much to the chagrin of the engineers it was found to perish rapidly and to be subject to attack by a peculiar fungus.

The expansion of trade and commerce in London needed an effective telegraph network to provide communication. In London, the problems of getting permission to put up overhead wires were multiplied; because of the multitude of cables and pipes already in the ground it was not considered practicable to lay underground cables. The sales operation occasionally misfired; the representatives of the telegraph companies seeking permission to run overhead wires took as specimens miniature telegraph posts, the size of a toothpick. So mysterious was the whole concept of sending messages by electricity that many householders were flabbergasted when workmen began erecting full size telegraph poles outside their back doors—they were expecting something the size of the original toothpick.

The telegraph system in London during the decade of private enterprise was never less than faintly ludicrous. Girls were advertised for to operate the transmitters and receivers, and to transcribe the messages, and the companies were overwhelmed by the response, and especially by the numbers of well-bred girls who were willing to take up this work for a mere ten shillings a week. The public were less enthusiastic, and at the end of the first year (1860) the 52 London stations had taken 73,480 messages, averaging seven messages a day at sixpence a time. The reason for the reluctance of the public to spend the equivalent of 20p (50c) on a telegraphic message was the extreme efficiency of the postal system; in the towns there were frequently six deliveries a day. The private companies were heartily glad when the Post Office took over the telegraphic system in 1870, for it never made more than a marginal profit.

The telegraph in the town brought with it the telegraphic exchange. The first patent for such an exchange was granted in 1851—'a particular combination of electric wires for the conveyance of intelligence in the interior of large towns whereby a central station has connected with it a certain number of houses in order that each house or subscriber may communicate privately with the central station'. The patent was more applicable to the later telephone than the telegraph, and anticipated the 999 emergency call system of our own day.

For a long time, the telegraph was the only medium making use of electricity. Lighting was possible, but it was arc lighting, using two carbons, much too bright and expensive for domestic use, or even for street lighting. An early use of the arc light was in the construction of the new Westminster Bridge in 1858; the most brilliant artificial light produced to 1858–9 was successfully tried at the South Foreland lighthouse, Dover, but it was not until 1878 that electric light became a commercial proposition, when it was used to illuminate the Gaiety Theatre and the Houses of Parliament. The incandescent lamp, the basis of electric lighting as we know it today, was held up by the difficulty of producing a vacuum in a glass bulb, but in 1879 this was fully solved, and by the end of 1880 Edison lamps, known as 'burners', were being manufactured in quantity.

The electric telegraph triumphed where other electric devices failed or were yet to be perfected because it needed only a small quantity of current. Electricity in the service of communication did not power anything; it only informed. The telegraph was significant not only for what it did, but also for for its side-effects. It showed how science could be integrated into the economy of a country, and it provided all the stock in trade necessary for electrical experimentation—batteries, terminals, insulated wire, coils, switches, measuring instruments. The telegraph has also provided a wealth of gadgetry, interesting today if hardly

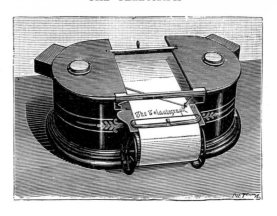

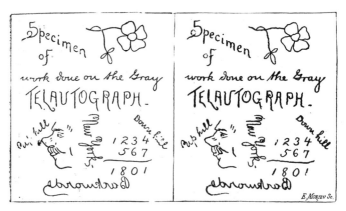

34 *The telautograph was more of a toy than a telegraphic instrument, and little was heard of it after its invention in the 1880s.*

usable, which, from the Wheatstone and Cooke five-needle telegraph onwards can certainly be termed true industrial antiques.

For those with large houses and a more than usual interest in gadgetry there is no reason why an electric telegraph could not be rigged up indoors, though it must be admitted, spectacularly visual as the equipment is, that it does not vie with an ordinary telephonic inter-communication system for efficiency. There is more telegraphic equipment about than one can imagine, and one can find the oddest things languishing in junk shop corners simply because no one knows what they were used for. An excellent place to seek out telegraphic equipment is anywhere where old railway equipment is being sold, and a visit to Kings Cross, where British Railways have a shop devoted to railway relics, might prove profitable.

CHAPTER 4

THE TELEPHONE

To MANY social historians, the telephone is the evil genius of communications, replacing the letter as the medium for exchanging trivia, the sort of trivia one finds in the letters of Jane Welsh Carlyle or Charles Lamb, to whom the written letter was an art form. The telephone is the instrument of instant decision, and has contributed more than can possibly be realised to the changes that have swept the world over the last half-century. In 1907 only one person in a hundred had a telephone; in the whole of London there were just 41,236 instruments. In Great Britain there are now more than ten million telephones.

Telephones have been in the forefront of industrial antiques for a number of years, for when it was realised that old telephones were quaint and could be rigged up to contemporary wiring, there was such a demand for them that several enterprising firms with an eye to the main chance began reproducing antique telephones, and unquestionably many people own these copies under the belief that they have genuine old telephones.

It is often believed that Graham Bell was responsible for the invention of the telephone, but pride of place must go to Philipp Reis, who began his researches in 1860. Reis was a German schoolmaster who had all his adult life been interested in the telegraph; his aim in life was to make the 'speaking telegraph'. 'I succeeded', he wrote, 'in inventing an apparatus that enables me to convert audible sounds into visible signs, and with which, moreover, sounds of every sort may be reproduced by the galvanic current at any distance. I called it the "Telephone".'

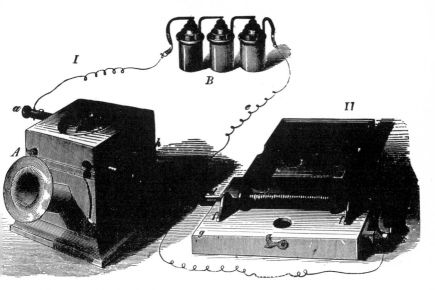

35 *Diagram of Reis telephone.*

Reis used a variety of objects for his experiments, including a wooden replica of a human ear, and a violin, and in 1861 he was sufficiently confident of his discovery to give a lecture to the Frankfurt Physical Association at which he enunciated the first principles of telephony :

> First—every sound and every combination of sounds produces in our ear vibrations of the membrane, the course of which can be represented by a curve. Secondly—it is the course of these vibrations alone that produces within us the sense of sound; thus every change of course must change the sense of sound. If, therefore, we can produce anywhere, by any method, vibrations similar to those of a particular sound or combination of sounds, we shall have the same impression which the sound or combination of sounds would have made directly on us.

To his audience, Reis's telephone was of no more consequence than the toy known as the string telephone—two cans connected by string—which was then popular. The net result of Reis's experiments was an order for a dozen telephones, which were made by a Frankfurt mechanic and sold at eight to fourteen thalers (£1.20–£2.10; $3.10–$5.45) each, and later a newspaper

75

article which described the Reis telephone as 'a child's toy for clever boys to make themselves'. He was invited to show his apparatus at another scientific junket in 1864; this, he thought, would finally put his telephone on the map. He was disappointed, and by the time he died in 1874 at the age of forty there was no indication that his telephone was, or could be, anything more than a toy.

The pattern of the sewing machine and the typewriter can be traced again with the telephone. Early work done by European

36 *The first Bell telephone of 1876, primitive, stark, and later to be hidden away in a box.*

inventors was unappreciated and ideas languished until taken up by Americans. Once more it was a question of economic necessity. When a device was *vital* to the success of an enterprise, the success story started in Europe and was followed or paralleled in America—as in the case of the electric telegraph, which was essential to the efficient working of the British railway network. Those who could have promoted the cause of the telephone in Europe were abysmally short-sighted; they saw it as a drawing-room amusement.

The energy and persistence of the Americans in bringing into being non-starters has irritated many historians of electricity, such as Silvanus Thompson (1851–1916) who championed Reis to the extent of writing a book about him, putting him in the forefront of telephone evolution. An article in 1914 in *The Times* ignoring the contribution of Reis drove Thompson into a paroxysm of fury.

Reis's telephone could transmit the pitch of a sound but not the quality. It could, at a pinch, carry a tune, but not a spoken sentence. Its main importance was that later it was used as a weapon to try to break patent rights.

To some extent the special pleading for Reis has overshadowed the very real contribution of Hermann von Helmholtz (1821–94), as it usually does when an amateur appears to outshine a professional. Helmholtz was one of the great scientists of the century, inventor of the ophthalmoscope in 1851; he also discovered the theory of colour-blindness, but perhaps his greatest work was done on the quality of sound. His pupil Heinrich Hertz contributed much towards the concept of wireless telegraphy. Helmholtz discovered that electromagnets kept tuning forks vibrating; by blending tones of forks he could achieve a simulacrum of the human voice, and he invented an apparatus for producing artificially the vowel sounds of the human larynx.

Alexander Graham Bell was a specialist working along the same lines as Helmholtz whose work he knew. Born and educated in Edinburgh, in 1870 Bell went to Canada, and thence to the United States, where he became professor of vocal physiology in the University of Boston, becoming involved in the rehabilitation of deaf-mutes. To enable them to see the effects of the sound-waves of speech, he set up two magnets, winding insulated wire round each, and connecting the coils by more wire.

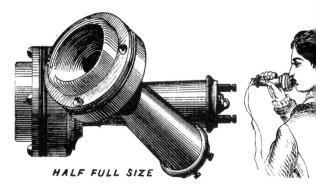

HALF FULL SIZE

THE TELEPHONE.

The Articulating or Speaking Telephone of Professor Alexander Graham Bell has now reached a point of simplicity, perfection, and reliability such as give it undoubted pre-eminence over all other means for telegraphic communication. Its employment necessitates no skilled labour, no technical education, and no special attention on the part of any one individual. Persons using it can converse miles apart, in precisely the same manner as though they were in the same room. It needs but a wire between the two points of communication, though ten or twenty miles apart, with a Telephone or a pair of Telephones—one to receive, the other to transmit, the sound of the voice—to hold communication in any language. It conveys the quality of the voice so that the person speaking can be recognised at the other end of the line. It can be used for any purpose and in any position—for mines, marine exploration, military evolutions, and numerous other purposes other than the hitherto recognised field for Telegraphy; between the manufacturer's office and his factory; between all large commercial houses and their branches; between central and branch banks; in ship-building yards, and factories of every description; in fact wherever conversation is required between the principal and

[1] P.T.O.

37 Advertisement for the Bell telephone of 1877.

In front of the first magnet he placed an iron reed, close to the second another. When he spoke loudly near the first reed it vibrated and affected the magnet, which sent electrical impulses along the wire to the second magnet, which caused the second reed to vibrate in the manner of the first.

Bell was a medical man venturing into the world of electrics. He knew nothing about electrical theory, but he did know about the make up of the human ear. He speculated that 'if the tiny disc within the ear can vibrate a small bone, surely vibrations of a thin iron disc can in some way be transmitted electrically to another iron disc. Why cannot the second iron disc be made to reproduce the original words to which the first iron disc had vibrated?' He obtained the ear of a dead man, attached a light straw to the back of the ear-drum, and rested the other end of the straw against a piece of smoked glass. By speaking into the ear the other end of the straw initiated a series of marks on the smoked glass.

Bell had abandoned his professorship, his money was running out, but in 1876 he succeeded in transmitting a message to his assistant in the basement from the attic. The primitive equipment he patented was described as 'an improvement in telegraphy'. On the same day another American, Elisha Gray, patented a similar contrivance.

The telephone was exhibited at the Philadelphia Exposition of the same year, where it was forced to compete with the first electric light, a telegraph printer, and Gray's invention. It was ignored until the arrival of the Emperor of Brazil, who took a great liking to it and focused the attention of the scientists on this unspectacular apparatus. The realisation was dawning that here was something of consequence, and the *Scientific American,* an organ of great importance in the circulation of novelties, covered the telephone in a full page spread.

Very shrewdly, Bell had patented his invention in Britain but had omitted to cover Germany. The firm of Siemens began manufacture of the telephone on a large scale, but it did not catch on, as it was only possible to speak from one room to another in the same house, and not until 1881 was a recognisable telephone service inaugurated and the first Berlin telephone directory published, with forty-eight subscribers.

The British were sharper. In 1877, W. H. Preece brought

38   *One of the first telephones seen in Britain, dated 1878.*

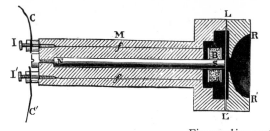

Fig. 302*d* is a section of the instru-
ment where N S is a cylindrical steel magnet, on one end of which is
wound the small coil B, made of fine silk covered copper wire, the extrem-
ities of which pass through the handle M at *f f,* and are connected by the
binding screws *I I'* with the line wire C C'. Close to the coil covered
end of the magnet is a very thin diaphragm of iron, L L', and when this is
thrown into vibration by the voice speaking into the trumpet-mouth
opening, R R', its movements produce currents in the coil according to the
principles that have already been explained, for it will be observed that
the iron disc is magnetized by the inductive action of the permanent
magnet N S. These currents passing through the coil of the receiving
instrument raise or lower the intensity of the magnetic force in it, so that
the distant disc reproduces the vibrations of the transmitter.

*39   A diagram of the Bell telephone. This served as receiver and
transmitter.*

the Bell telephone to England, and it was shown off at the
Plymouth meeting of the British Association, followed by wide
demonstrations in London and elsewhere. It was not received
with universal warmth. The *Saturday Review* said that it was
little better than a toy; 'it amazes ignorant people for a moment,
but it is inferior to the well-established system of air-tubes.'
However, the Earl of Caithness was impressed. The telephone
was the most extraordinary thing he had seen in his life. And
so was Queen Victoria, who was confronted with the telephone
in 1878. There is some doubt where this occurred—at a London
private house, at Buckingham Palace, or at Osborne in the Isle
of Wight, but the approval of the Queen put the seal of success
on the telephone in England, the Telephone Co Ltd was set up
with a capital of £100,000, and in August 1879 the first telephone exchange was opened in Britain, at 36 Coleman Street,
London, catering for 7 to 8 subscribers (though 150 could be ac-
commodated). By 1885 there were 3800 subscribers in London,
and 10,000 in the rest of the United Kingdom, a modest success
when at the same time the Bell Telephone Company of America
had 134,000 subscribers.

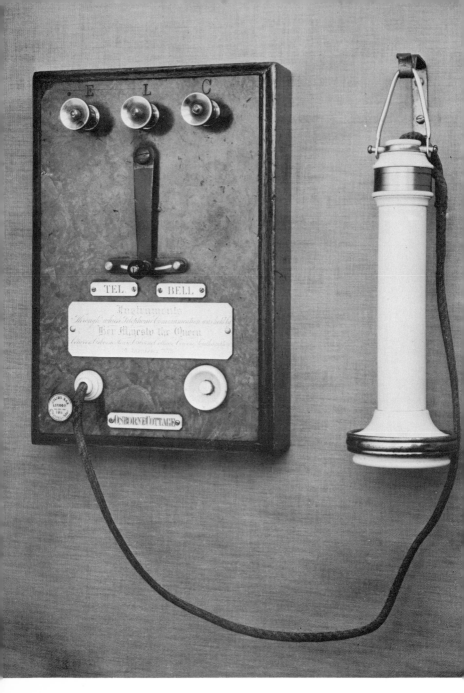

40   The instrument presented to the Queen in 1878.

In the early day the transmitter and the receiver were the same. A circular object poked out of a box, and a sign by the telephone read: 'Don't Talk with your Ear or Listen with your Mouth.' The provision of separate transmitters and receivers was a logical development which was followed in due course by having the transmitter and receiver in one unit, as in the telephone of today.

The design of telephones followed the trends. The instruments of the 1880s were functional, primitive, and ugly, but by the turn of the century, when most of the technical problems had been solved, they became ornate and decorative, with art nouveau ornamentation proving eminently suited to the apparatus. The telephone of 1920 was spare but attractive, with its slim stem and mixture of black and chrome; indeed, it was one of the few representatives of fitness for purpose that circulated in the years that immediately followed the 1914–18 war. The sheer nakedness of the telephone was repellent to the sensitive. It was necessary to disguise it as something else, or to hide it. This yearning had not died out with the Victorians, as is evident from the strictures of Aymer Vallance writing in the *Magazine of Art* 1904:

> The perverse ingenuity which turns drain-pipes, set up on end, into vases for bulrushes [sic] and dried grasses; milking stools into flower-stands; cauldrons into coal-scuttles; whereby square pianos, disembowelled of their works, become 'silver tables,' and sedan chairs, fitted with shelves, become china cabinets, is truly marvellous and worthy of being expended in a better cause.

Unlike other nineteenth-century inventions that changed the world, such as the typewriter, the telephone was not susceptible to any great changes, only improvements. The major improvement was the carbon transmitter of Thomas Edison in place of the electromagnetic one used by Bell; voices could be heard much louder.

The monopoly of the American telephone in Britain saved Britain from the patent wars that were a consequence of most inventions. The Western Union had ignored the telephone, but when one of its subsidiary companies reported that several of its machines had been ousted by the telephone they organised the American Speaking-Telephone Company, and, co-opting Edison

41   Diagram of a wall telephone of 1878.

and two other electrical inventors, attempted to swamp Bell. The patent war lasted 11 years and comprised 600 lawsuits. Bell and the Western Union reached an amicable agreement, but other companies and individuals persisted in trying to break Bell's 1876 patent, without success.

The telephone shares with the telegraph the distinction of using very small quantities of electricity. There were no difficulties in generating sufficient current for the telephone. As Herbert Casson's lively but not altogether accurate *The History of the Telephone* (1917) has it

> Cool a spoonful of hot water just one degree, and the energy set free by the cooling will operate a telephone for ten thousand years. Catch the falling tear-drop of a child, and there will be sufficient water-power to carry a spoken message from one city to another.

It was difficult to present the telephone in terms of something else, so there was no choice but to hide it. In 1919, the column 'Things you want to know' in the new magazine *Our Homes and Gardens* answered a question on telephones:

42   *An early telephone with separate transmitter and receiver.*

> Admittedly the telephone is a distinctly unbeautiful instrument,
> and your desire to conceal it is not uncommon. The instrument
> can be very effectively accommodated in a recess in the wall,
> covered by a panel . . .

The magazine returned to the matter a year later in a full length
article entitled 'Camouflaging the Telephone'.

> The telephone has provided yet another problem for those of
> us who wish to have nothing that disturbs the decorative effect
> of our rooms; and as the design of the telephone is purely
> mechanical, it does not fit happily into the scheme and we
> desire to camouflage it in some manner.

The favourite method was the telephone box, looking like nothing
so much as a biscuit tin. Pseudo-Japanese styles were very much
in vogue, and the telephone boxes were heavily lacquered in that
manner. A curio of this type was the commercially produced
'Chinese Chippendale Telephone-box' with a carved pagoda top
and finished in black and gold. There were also ornate pieces of
furniture resembling those pot-cupboards found in old fashioned
hotels, concealing not the chamber-pot but the telephone. In
bedrooms it might be inconvenient to have a box overcrowding
bedside table space. 'In such cases we must either make up our
minds to face the naked mechanism or make a small recess in
the wall.' Or, better still, hide it on a bookshelf behind a screen
of dummy book-backs.

It is evident from these extracts that the owners of telephones
were ashamed of them, that they had not come to terms with
technology in the home. It is something of a paradox that the
modernist movement that started in the mid-1920s, and became
accepted by the readers of such magazines as *Our Homes and
Gardens* in the early 1930s, left the telephone high and dry. In
the austere geometric homes of the 'thirties the telephone was
still the 'naked mechanism' of 1920, apparently not amenable
to the streamlining that affected furniture or electric light fittings.

The economic and social effects of the telephone are apparent.
The telephone also contributed to technological knowledge,
stimulating inventors into venturing into new territory. The
rivalry between Graham Bell and Thomas Edison put each
on his mettle, and just as Edison improved Bell's telephone trans-

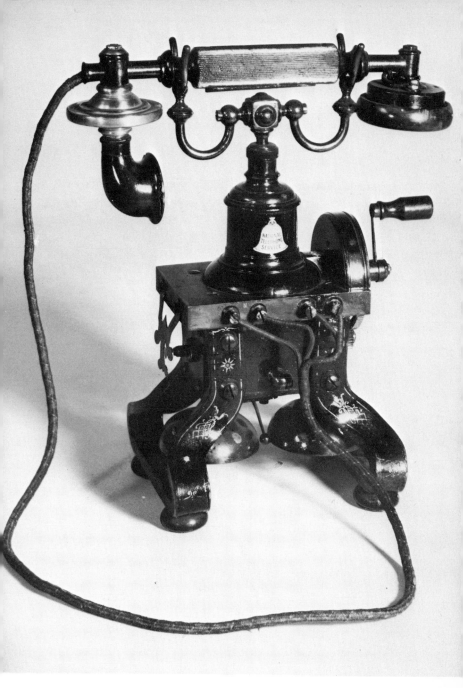

43 *The transmitter and receiver incorporated in a hand-held unit.*

mitter or microphone so did Bell try to supersede Edison's phonograph with his own graphophone of 1887. Specialists in telephony had been blooded in telegraphy (Edison started his working life as a telegraph operator), and technical problems common to both telephony and telegraphy had been solved by telegraphic engineers for the benefit of all, such as overhead wiring versus underground cables.

Bell related in later life that had he had any knowledge of electricity he would never have ventured into the world of the telephone, but other inventors were not so modest. If Bell, visualised as an innocent amongst the experts, could come up with such a marvellous instrument, could not they make something even more wonderful? One product of this mode of thought was the water-jet telephone transmitter, in which the power was supplied by water, a short-lived and obscure invention. In 1873, Willoughby Smith discovered that when selenium was exposed to light its electrical resistance varied with the intensity of the light falling upon it; selenium under the influence of light, especially the sun, acted as a small battery. Without realising it, Smith was taking the first faltering steps towards solar power, only now utilised in space flight.

From this discovery evolved the photophone, constructed by Bell and Sumner Tainter of Washington in 1880. A thin plane mirror is thrown into vibration by the voice, a beam of light is reflected from this mirror and received at a distance by a cell of selenium, connected with a telephone, which reproduces the sound. The specification 'at a distance' must be viewed sceptically, and the photophone is interesting not for what it was but for the germ of the idea that initiated it.

The adaption of 'antique' telephones to modern use is not difficult, though it is not likely that the earliest telephones, in which the transmitter and the receiver is one and the same, would be considered an asset in the modern home, quaint as the exhortation not to listen with the mouth or talk with the ear undoubtedly is. It is not likely that instruments suitable for converting on aesthetic grounds will have the Bell electromagnetic transmitter and not the improved carbon transmitter. But as anyone who has accidentally unscrewed the dial or ear-piece of a modern telephone will have discovered, the actual components of a telephone are ridiculously small in number, and where an attractive turn of the century telephone is bought with the inner workings in disarray, a telephone engineer will have no problem in transplanting a modern replacement.

CHAPTER 5

THE CAMERA

THE SEWING MACHINE, the typewriter, these enabled the users to do things more quickly. The telegraph and the telephone meant instant communication. All were important and all revolution-ised activity, but in human terms the camera was far more sig-nificant, for it helped people to *see* differently and to see object-ively. For the first time, strange objects and strange scenes could be viewed as they really were, without the necessary filter of an interpreter. Exact and precise as were the topographical artists of the early nineteenth century, the presence of a human element prevented the exact replica of the subject.

The philosophy of the sewing machine and the typewriter was put succinctly by William Morris in his *The Aims of Art* (1887).

> Machinery will go on developing, with the purpose of saving men labour, till the mass of the people attain real leisure enough to be able to appreciate the pleasure of life; till, in fact, they have attained such mastery over Nature that they no longer fear starvation as a penalty for not working more than enough . . . They would soon find out that the less work they did, the more desirable a dwelling-place the earth would be.

The sewing machine and the typewriter were aids to existence, labour-saving devices, leaving needlewomen and clerks time to pursue more life-enhancing projects. That it has not worked out like this was not lack of imagination on the part of William Morris or John Ruskin or the sages of the period, but their over-estimation of the desire of the masses for improvement.

The camera belongs to a different category, an instrument for greater knowledge and awareness. This was not realised when it first appeared. Ruskin saw it as an accessory for artists, providing the equivalent of a preliminary sketch. The painter Paul Delaroche was more acute; upon the publication in 1839 of the first practicable process of photography he exclaimed 'From today, painting is dead!' though this did not prevent him continuing to paint massive pictures such as *Marie Antoinette before the Tribunal* (1851) or *The Finding of Moses* (1852).

It is difficult to say why photography was not invented before it was, for it did not depend on technological know-how or the practical use of electricity. The optical and chemical principles on which photography is based were well known in the eighteenth century, and the *camera obscura*, the direct ancestor of the photographic camera, was known nearly a thousand years ago. The *camera obscura* was a dark room with a small hole in the wall through which the view outside was projected upside-down on to the facing wall. Brighter sharper images were created by the use of a lens in the hole (1550) and the reversal of the image was corrected by the interposition of a concave glass (1573). The portable *camera obscura*, a hut between poles, was promulgated in 1646, followed by the first box camera in 1657.

The provision of a plane mirror set at 45° to the axis of the lens created the first reflex camera, with the image the right way up, in 1676, and less than ten years later the oiled-paper screen used in the 1676 camera was replaced by a ground-glass screen (still used today in reflex cameras).

The first device to use a simple lens systematically was the magic lantern or lanthorn, first promulgated in *Ars magna lucis et umbrae* (1646) by Athanasius Kircher. The magic lantern was first used to show comic pictures, and was utilised by wizards and magicians in the creation of 'ghosts', but in the nineteenth century it came into its own as an instrument of education, revealing the mysteries of small organisms and the glamour of far-away places.

In early lanterns only a single plano-convex or bull's eye lens was used, and as a light source was also a heat source lanterns had to be constructed so that the heat would not crack the slide. Before the days of photography, subjects were painted or transferred on to glass slides. As educational devices, magic lanterns became increasingly sophisticated, and were popular well into the present century with an electric bulb instead of a naked light.

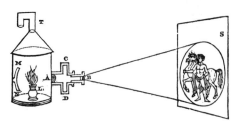

44   Diagram of the magic lantern.

A favourite entertainment for Victorians were dissolving views, provided by two magic lanterns, side by side or one on top of another. The fronts of the lanterns were slightly inclined to each other to ensure that the illuminated discs on the screen coincided. Thin metal shutters culminating in comb-like teeth cut off the supply of light of either lantern at will, and in this way views and subjects gave the impression of melting or dissolving. The dissolving view dates from about 1811.

Contemporary with the dissolving view was the phantasmagoria, in which the screen was set between the lantern and the audience. The screen was semi-transparent, and as the lantern was rolled on wheels nearer to and further from the screen so did the picture enlarge or contract. An automatic device kept the picture in focus.

Before the end of the seventeenth century all the optical elements existed for the photographic camera. The only problem was the retention of the image. In 1772 observations on the effect of sunlight on silver salts were published, confirming what had been known by Fabricius, who had stated his findings in *De Rebus Metallicus* (1566). In 1774 Dr Hooper in *Rational Recreations* described a method of writing on glass by the rays of the sun, using a mixture of chalk, acid, and silver salts. In 1802 Thomas Wedgwood and Sir Humphrey Davy published in the journal of the Royal Institution 'An Account of a Method of Copying Paintings upon Glass, and of making Profiles by the Agency of Light upon Nitrate of Silver'.

All the chemical elements were therefore there, and there

seems grounds for believing that some time during the first quarter of the nineteenth century someone actually succeeded in fixing an image. In 1871 there was an intriguing account of a silvered copperplate being discovered in the lumber-room of Matthew Boulton's house in Birmingham, depicting the Boulton and Watt factory at Soho, Birmingham, the first modern industrial plant in the world. And Boulton had died in 1809.

Traditionally the honour of first permanently fixing a photographic image goes to Joseph Niépce (1765–1833), using bitumen of Judea, which hardens instead of changing colour under the influence of light. The world's first photograph (1826) is a view from his window; exposure time was eight hours. Niépce also invented the bellows-camera, and the iris diaphragm.

After failing to interest the Royal Society in his invention, Niépce went into partnership with L. J. M. Daguerre in 1829. Daguerre was a theatrical designer, inventor of the diorama, an exhibition of pictures, illuminated and viewed through an opening in the wall of a darkened chamber (now a standard feature of enterprising museums). By using a system of development with mercury vapour, Daguerre cut down the exposure time from several hours to a matter of 20–30 minutes. The 'daguerreotype', a photograph on copper, was born. Daguerreotypes, often mounted in attractive wallets, are still about in great quantities, and because of their historical significance it is worth while describing the process of taking a daguerreotype in full.

A polished silvered copper plate was exposed to the vapour of iodine, causing a layer of light-sensitive iodide to form on the surface. The plate was exposed, then developed in a box containing mercury, which was heated to vaporise it. During this process minute globules of mercury settled on those parts of the plate that had been affected by light (the lights of the picture). Those parts unaffected, the unchanged iodide, were dissolved away by washing with thiosulphate alias sodium hyposulphite ('hypo'). The image was now permanent, but the mercury only adhered lightly, and the daguerreotype needed to be protected by glass.

In 1839 the French government acquired the process by act of Parliament, awarding the inventors pensions. Two daguerreotype establishments were formed in London armed with full patent rights, but their proprietors only once enforced these rights, and the way was clear to anyone who cared to venture into this field.

At first this field was limited. Portraits were not possible because of the long exposures needed. In the jury reports of the Great Exhibition of 1851 it is stated that to procure a daguerreotype portrait it was 'required that a person should sit without moving for twenty-five minutes in a glaring sunshine'. However, the jury was out of date. Alexander Wolcott of New York devised a camera without a lens, the image of the sitter being reflected on to a sensitive plate by a concave mirror, and although this image was small (two inches square) and not very sharp, Wolcott opened the world's first portrait studio in 1840, and a year later London followed suit. The only advantage of Wolcott's camera was that exposure times were short.

More significant was the portrait lens introduced by J. F. Voigtländer in 1840. Exposure times were cut to seconds rather than minutes. This lens was so successful that it was not superseded until 1889. Coincident with the introduction of this lens was the speeding up of the chemical process, sensitising the daguerreotype plate with vapours of bromide or chlorine as well as of iodine. With exposure times of between ten and ninety

45 *Elegant and simple, this daguerreotype camera of 1840 was the basic design of camera until the coming of roll film in 1889.*

seconds, photographic portraiture became a craze.

The disadvantages of the daguerreotype were that it was not cheap, and only one could be made. Nevertheless the complete photographic outfits sold by Daguerre, termed 'official' and bearing his signature, were popular, and perhaps there are still complete outfits in existence. The outfit included plate box, iodising box, mercury box, spirit lamp, chemicals in earthenware jars, powders, polishing cloths, etc. The camera consisted of two boxes, sliding one within the other, one containing the ground-glass screen for focusing, the other the lens. This was the standard camera of the period, though the bellows was to supersede the primitive sliding action of the two boxes.

At the same time as Daguerre was perfecting his processing in France, W. H. Fox Talbot was experimenting with taking photographs on paper coated with silver chloride. The daguerreotype was announced on 7 January 1839; Talbot hurriedly exhibited his 'photogenic drawings', and in 1841 he patented his 'calotype' process. The calotype lacked the fine detail of the daguerreotype, detail as good as that on fine grain film of today, but was suitable for views, where broad effects were more important than minutiae.

However, the success of photography depended on the response of the public, more interested in having their likenesses taken than in acquiring views. The daguerreotype flourished until the late 1850s, but both photography on copper and on paper were rendered obsolete in 1851 by photography on glass, which held the field for more than thirty years. Plate cameras are still used by professional photographers.

The wet collodion process was messy; collodion is gun-cotton dissolved in ether. It was poured over a glass plate, and tilted until the solution formed a sticky covering. A reporter of *Household Words* in 1853 watched this in awe: 'To do this by a few left-handed movements without causing any ripple upon the collodion adhering to the glass is really very difficult'. The plate was then sensitised by dipping it into a silver nitrate solution. The exposure had to be made while the plate was still damp. Development and fixing were quite straightforward.

The collodion process had many advantages. More susceptible than previous processes to light and shade, the most insignificant detail could be drawn forward. Like the calotype but unlike the

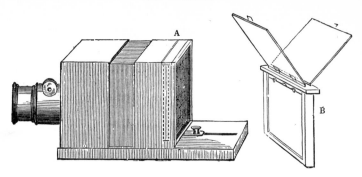

46   The standard plate camera. B is the plate in the hinged holder; the holder, when closed, slots into A.

daguerreotype, many positive prints could be made from the one negative. Calotypes faded badly; not so prints produced by the collodion process. But, if anything, cameras became bulkier. The paper on which the prints were produced was not very sensitive, and enlarging was little practised. Negatives were large, up to 16 x 12in, and the cameras had to be big enough to accommodate plates of this size. This did not matter much for studio photographers, but for travelling photographers the impedimenta needed was ridiculous. The photographic perambulator was no joke, for the total equipment could reach eight stone and the early folding cameras did little to cut down on this weight.

The Great Exhibition gave a considerable boost to photography, and particularly to that eccentric variety known as the stereoscopic; the first stereoscopic photographs were taken by two cameras mounted side by side, but the twin-lens camera came in 1853. These cameras cut down exposures to a fraction of a second. Moving subjects at a distance could be taken. Stereoscopic views became a craze, and viewers were produced by the tens of thousand. The two photographs are pasted side by side, and when viewed through the two eye-pieces they merge into one, giving a 3-D effect. These viewers and their photographs are to be found in large quantities, and it is interesting to note that many of them are of saucy subjects, paralleling the increasing market in pornographic photographs. The stereoscopic cameras are particularly interesting, as they are much smaller than the large plate cameras, and foretell the emergence of the miniature, though the pistol camera of 1856 and Dubroni's hand-camera of 1864 were, at those dates, exceptional. One of the holders of the daguerreotype franchise in Britain produced a variation of the stereoscope; the two photographs are taken with a slight time lapse, and by closing one eye and then the other, the impression of movement is produced. Two popular examples were a fist-fight and a needlewoman plying her trade.

One of the greatest commercial innovations was the rise of the *carte de visite* photograph, common from 1859 to the end of the century. These were often taken with a special camera with four identical lenses. Eight pictures were taken on one plate in two exposures, the contact print was cut up, and the portraits were

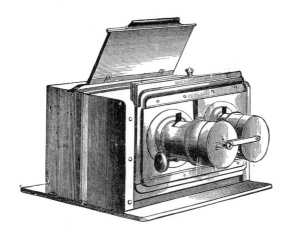

47   *The stereoscopic camera, first seen at the Great Exhibition of 1851.*

mounted on cards. There was a massive demand for portraits of royalty and celebrities, and between 1860 and 1862 3,500,000 cartes (the word became Anglicised very quickly) of Queen Victoria were sold. There arose a subsidiary industry—the production of photograph albums, often sumptuously bound in papier-maché or leather, with frequently a small musical movement inserted in the back cover (more often than not playing 'Home Sweet Home').

'Dry plate' photography was introduced in 1856, in 1871 gelatine emulsion was introduced, and by 1878 improvements had made gelatine dry plate photography (gelatine = nitric acid,

48 A studio camera of about 1890.

49   This curious 1887 contrivance is a telescopic camera.

cadmium bromide, and silver nitrate) twenty times faster than wet collodion. The way was clear for genuine instantaneous photography. For the first time a means was found to print photographs on to news print, and this brought photography to the notice of everyone. Glass, which had proved such a godsend since 1851, was now replaced by celluloid (nitro-cellulose and camphor), used initially for solid objects, and produced in America in sheets with a uniform thickness of a hundredth of an inch in 1888. It was still too thick for roll-film, but this was remedied the following year when the Eastman Company of America was granted a patent. The introduction of dry plate photography cut out much of the paraphernalia surrounding the craft. The Kodak camera of 1888 reduced the number of operations needed to make a photograph to three. Commercial developing and printing took all the work out of the amateur's hands.

With the perfection of the instrument came gadgetry and the 'detective' camera, the camera that looked like something else—binoculars, revolvers, books and watches, or concealed in hats, purses, beneath jackets. The candid camera appealed to the

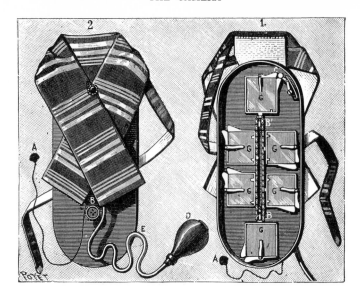

50 *This version of the detective camera was used to produce the facsimiles.*

mentality of the 'nineties, and it is rather a pity that these gimmicks, ingenious though indifferently made, have more appeal as interesting objects than the earlier cameras which, simple as the components are, are beautifully made, as much a product of the cabinetmaker as the optical manufacturer. The Voigtländer lenses of 1840 and after are splendid specimens of the craft, though it would be a pity if old cameras were cannibalised for the sake of their lenses.

The interest of early cameras has for some time been overshadowed by the attention that has been bestowed on the photo-

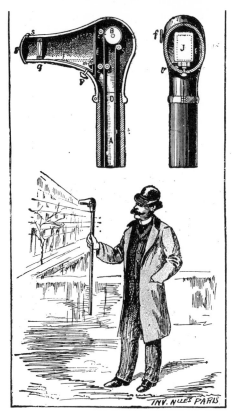

51 *The walking stick camera was an ingenious variation. A band
of sensitised celluloid passes round a pulley (B), which is turned
by a button (f). In order to take a picture the cap (r) is
opened in front of the lens (q), and by pressing the knob (p)
an aperture in the screen (E) is opened and the exposure
is made. The cap is then replaced, the button (f) is turned,
and another portion of the band is brought into position for
the next exposure.*

graphs that they produced. The photographs of Julia Cameron,
when signed and authenticated, can fetch well over a hundred
pounds each at auction, and the genre scenes of the 1860s are
highly sought after. Photography did not kill painting, though it
prompted it into new channels. For a long time photography
lacked colour. A London photographer, Mr Mayall, experi-

52 *The camera disguised as binoculars.*

53 *The same binocular case shown opened.*

THE KODAK

Is the smallest, lightest, and simplest of all Detective Cameras—for the ten operations necessary with most Cameras of this class to make one exposure, we have **only 3 simple** movements.

NO FOCUSSING.   *NO FINDER REQUIRED.*

Size 3¼ by 3¾ by 6½ inches.   **MAKES 100 EXPOSURES.**   Weight 35 ounces.

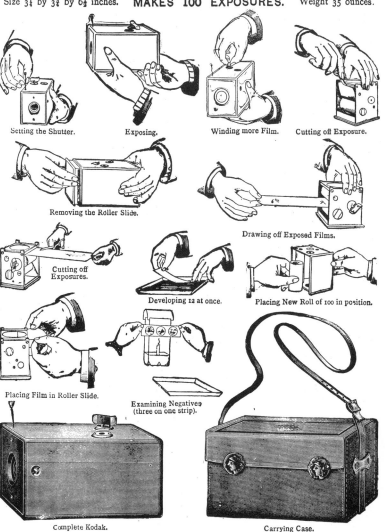

Setting the Shutter.   Exposing.   Winding more Film.   Cutting off Exposure.

Removing the Roller Slide.

Drawing off Exposed Films.

Cutting off Exposures.

Developing 12 at once.

Placing New Roll of 100 in position.

Placing Film in Roller Slide.

Examining Negatives
(three on one strip).

Complete Kodak.   Carrying Case.

FULL INFORMATION FURNISHED BY

THE EASTMAN DRY PLATE & FILM Co., 115, Oxford St., London, W.

54   An 1889 advertisement for the Kodak 'the smallest, lightest and simplest of all Detective Cameras'.

mented in 'crayon portraits in daguerreotype', which did not last long. Hand-tinting was usually crude, more because of the sitters' demands than lack of expertise. It was usually done by girls, whose pots of pigment were labelled 'sky' or 'flesh'. An observer, admiring the work of two of these girls, praised the subdued tones of a red coat. 'Yes,' said the girl, 'but I must make it redder presently; when we don't paint coats bright enough, people complain. They tell us that we make them look as if they wore old clothes.'

Photography influenced painting considerably. For such paintings as *The Railway Station* and *Derby Day* Frith used dozens of photographs; commissioned paintings of large gatherings, beloved of mayors and royalty, were based on innumerable small photographs pasted on to a master plan. Degas used photographs to see how exactly a horse looked in motion. Conversely, painting influenced photography. The type of picture immortalised by *When Did You Last See Your Father?* sparked off many photographers to re-create the same thing on glass instead of canvas. The camera was very good at reproducing pathos, and studies of distress and anguish were produced in tens of thousands.

Unquestionably magic lanterns are more sought after than old cameras, simply because they can be used to provide an entertainment. There is no way to show slides except by a magic lantern. Plate cameras, nevertheless, are by no means purely decorative, and although the processing is messier than is the case with modern roll film it is just as simple and newcomers to photography using glass plates will be astonished by the excellent detail obtained. Because of the lack of grain, plate photography is still used for portrait and newspaper work.

One should look for old cameras in photographic rather than antique shops. They are still being traded in part exchange for a modern camera. No doubt there will be a day when between-the-wars folding Kodaks will be sought after, but all cameras using roll film—except the very first—must be rated as relatively modern, and must therefore be judged on performance.

CHAPTER 6

THE CINEMATOGRAPH

It is apparent that the example of sewing machines and typewriters was again followed by photography—initiation and development in Europe, and exploitation in America. The first photographic studio was opened in America in 1840, and it was Eastman who put the camera within reach of the masses by making a simple camera with one stop, one speed, and one aperture, and by realising the possibilities of the roll film, which in turn was made possible by the American invention, celluloid.

The most cursory glance at the history of photography prompts the question: what were the scientists doing? Why was the development of the craft left to dilettantes and amateurs, or people from another sphere altogether—Daguerre, for example, a theatrical designer? Part of the answer lies, with a few honourable exceptions, in the inferior quality of the scientists compared with their illustrious predecessors, for they lacked vision and imagination, preferring to try and solve insoluble problems hallowed by tradition, or frittering away their energies in projects that were outside the scope of science.

The reasons why photography took so long to develop are obscure; in comparison, the history of cinematography is clear and simple. Cinematography had to wait for the arrival of flexible film before anything could be done, and when it did arrive, progress was rapid and sure.

In essence, cinematography is a sequence of separate images exposed at predetermined intervals representing successive phases

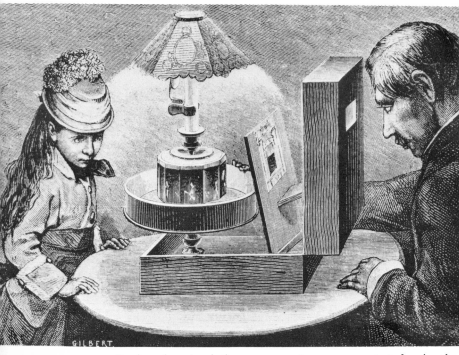

55 *Devices for simulating movement were very popular in the last quarter of the nineteenth century. The praxinoscope was more sophisticated than most; the picture on the rim of the wheel is reflected in a sequence of mirrors.*

of a movement. When exhibited in rapid sequence the images persist long enough to give the appearance of continuous motion. Cinematography is based on the phenomenon of persistence of vision; an image in the human retina will remain unchanged for a period of from one-twentieth to one-tenth of a second. This had been known theoretically since 130 AD, but the first person to use the principle was a British doctor, J. A. Paris, who invented a 'thaumatrope' in 1826, a trivial toy, consisting of a disc with a picture on either side, suspended between two lengths of silk. By turning the disc rapidly the two images mingle and appear as one.

More toys of this kind followed. Discs with slots in their rims were spun in front of a mirror; at the centre of the disc were drawings, each one slightly different from its neighbour. The reflection in the mirror viewed through the slots in the moving disc gave the impression of progression, and these toys were widely popular, whether they were called magic discs, fantascopes, phenakistiscopes, or stroboscopes. A more sophisticated version was called the zoetrope or wheel-of-life, in which a sequence of drawings was placed inside a horizontally-moving rimmed wheel, with vertical slots.

The magic lantern was used with moving slides, and throughout the 1860s many patents were taken out for projecting and viewing moving pictures, all relying on the principle of persistence of vision. These objects, simple as they are, are eminently collectable, and have the advantage that they can still be used, though only small children are likely to get excited by them. With the aid of lenses it was simple to project transparent photographs on to a wall or screen, and it was only a matter of time before all the elements of cinematography were correlated. The major step was to take a series of photographs, so that a consecu-

56   *The zoetrope, a toy with a revolving picture strip gave the appearance of movement and preceded the cinematograph.*

tive movement could be seen as a succession of single shots, and there came a convenient prompting from science. Due on December 8 1874 was the transit of Venus across the sun. The director of a Paris observatory placed a circular disc covered with photographic emulsion behind a second disc with slits. The discs were rotated by a clockwork movement, and the whole apparatus was attached to the eye-piece of an astronomical telescope. This was probably the first movie made; the movement of Venus across the sun was recorded in 48 consecutive pictures.

Astronomy had been one of the first sciences to use photography. As early as 1840 a Dr Draper had succeeded in daguerreotyping the moon, the first attempts to photograph the sun were made in 1845, and in 1860 a solar eclipse was put on film, especially interesting to astronomers because for the first time the flames surrounding the sun could be seen to be part of the sun itself.

In 1872 E. Muybridge was asked to record the individual paces of a galloping horse, and by using a series of cameras placed a foot apart, and triggered by electromagnetic means, he not only recorded horses galloping, but men, women, children and animals engaged in a wide variety of movements. These pictures, 20,000 in all, were of great interest to artists and scientists, and although he did not contribute much to the development of cinematography it is difficult not to admire his diligence and perseverance.

More important was E. J. Marey (1830–1904). He started his work in 1882, first of all using the principle that had proved so successful in photographing the transit of Venus across the sun. Marey benefited from the progress made in shutter mechanism, for his exposure times were 1/720 sec. A glass disc received 12 consecutive images a second; for each shot, the disc was stopped behind the shutter. Although modestly effective, this system had drawbacks, but by 1887 lengths of photographic paper had become available, and Marey constructed the first movie camera, his *chambre chronophotographique*.

Initially he made the mistake of running the film continuously past the camera lens, but soon realised that to recapture movement it was essential that for the exposure period the film must be halted in front of the lens. The clue to the successful camera was intermittent motion. Several of these cameras were made in

57   *The mutoscope, a very collectable device simulating movement.*

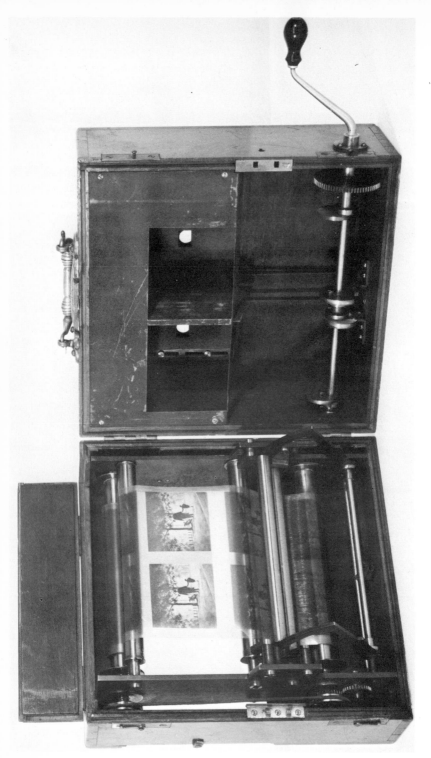

58 *A stereo camera by W. Frieze-Greene anticipating his ventures into cinematography.*

France to Marey's specification, but there are only two known to be in existence. All the elements of the modern movie camera are there, though the film was driven round by a hand-crank rather than by clockwork or an electric battery. A vital feature was the loop of film each side of the lens, to prevent the paper breaking.

With the coming of flexible photographic material the way was clear, and although the pioneer work had been done in France, other countries were soon busy. In 1889 Edison visited Paris and was impressed by Marey's camera, proceeding to evolve his own kind of intermittent movement in his kinetograph. In England, W. Friese-Greene and others demonstrated their movie camera to the Bath Photographic Society in 1890. Friese-Greene had begun by using paper soaked in castor oil to make it transparent, but he was one of the first to realise the possibilities of celluloid, and processed it himself, clarifying it and bringing it to the requisite thinness by rolling it through a household mangle. His camera was in two compartments, one above the other; by turning a handle the film in the upper box was exposed behind the lens, and moved into the lower box. His first movies were London traffic scenes, and a lady with a boy in Hyde Park. The journal *The Optical Magical Lantern* reported : 'It would doubtless seem strange if upon a screen a portrait of a person was projected and this picture slowly became an animated character, opened its mouth and began to talk. . . . Strange as this may seem, it is now an accomplished fact.' Although the mood of Victorian England was changing, Friese-Greene did not find backers for his camera, and turned his attention to colour film, X-rays, and the wireless.

The Americans were rather more enthusiastic. Edison made a film using 158 single shots, which received a good deal of attention, and the Kinetoscope Parlor was opened on Broadway, New York, in 1894. These were heady days for the industry, and there must be a good deal of neglected apparatus about, unconsidered and unrecognised, whether it is the 1893 bioscope of Marey's assistant Demeny, Edison's kinetoscope, or Anschütz's tachyscope of 1889 (patented in 1897). Of these, the most efficient was Edison's. Edison was one of the first to use perforations on photographic film, vital for a regular intermittent movement, and to use film 35mm wide, still the standard gauge.

59   Edison's kinetoscope was a late Victorian 'What the Butler
Saw'. The intricate array of wheels is to transport the film strip.

Once a reliable camera was produced, it was a simple matter to make a good projector. The optics were child's play, when one considers the long existence of the magic lantern. Edison's projector used electricity both for the light source and the film transmission.

The cinematographic trade started in England in 1894, when two Greek showmen brought with them a kinetoscope which they set up in a shop near Liverpool Street Station. The basic kinetoscope was on the lines of 'What the Butler Saw' in old-fashioned amusement arcades; for twopence a viewer peeped through a slit and turned a handle. Robert Paul, an instrument-maker in Hatton Garden, was approached to make more kinetoscopes, which he did, Edison having failed to take out an English patent. Edison refused to sell film to Paul, and Paul was forced to make his own. In 1896 he demonstrated his moving pictures at Finsbury College, the Alhambra music hall took a ten minutes short, 'The Soldier's Courtship', and the first British newsreel shots were taken at the Derby of 1896.

Edison was also the spur to the French cinematograph. Auguste and Louis Lumière took as their initial model the kinetoscope, but refined it, engaging C. Moissant to design for their camera a claw-movement which allowed two pins to enter the perforations at the side of the film, and pull it down while a semi-circular shutter cut off the light between lens and film. When the picture was exposed in the camera, or projected, the two pins of the claw-movement returned to their original position in time for the repeat performance. This technique became standard practice. The Lumières patented their camera and projector in France and England in 1895, and made a number of films including 'The Baby's Meal' and 'The Arrival of a Train' which played to audiences of 2000 a night. The Lumières' equipment was greeted enthusiastically, and in 1896 the London Polytechnic invited them to give a series of half-hour shows, accompanied by noises off provided by the Polytechnic staff. The music halls competed for this novelty, and the 'jerking screen' was the talking point of 1896.

The talking picture was well in the future, but as early as 1880 an American, Charles E. Fritts, had taken out a patent for recording sound by photography. Edison had got into a rut by attempting to couple his kinetoscope with a phonograph, calling it a kinetophone. In 1902 Léon Gaumont showed three films with dialogue synchronised on a phonograph, and a year later Oskar Messter invented the chronophone, the sound being produced by compressed-air loudspeakers. The projection cabin was equipped with a dial to help the operator synchronise the sound with the movements on the screen.

Other attempts to bring sound movies were the synchronoscope of 1908, and the graphophonoscope, while the first man to take out a patent for the simultaneous recording of picture and sound *on film* was an English electrical engineer, Eugene Lauste, in 1906. In 1920 Theodore Case patented his photo-electric cell system called movietone, followed by phonofilm in 1923, and

60 *Robert Paul's cine camera of 1896.*

61 *Thomas Edison was concerned with telegraphy, telephony, the
phonograph and the cinematograph. This caricature appeared
in Britain in 1901.*

vitaphone in 1926, leading inevitably to the first talking film, 'The Jazz Singer', in 1928.

Because of its weight and bulk it is doubtful whether the equipment of this century is collectable, though certainly that of the pioneers is eminently so. The question is whether such equipment can be put in working order, and whether the electric motors of that period can be keyed in to present day voltages. A period projector that works is decidedly a novelty, but, again, there is a snag. The sizes of film used varied enormously, from 17.4mm to 70mm, and so much did the perforations vary that every camera bought was provided with a perforating machine. The Lumières used one perforation per frame; their film was projected at 16 frames per second (a norm for silent films).

One of the few elderly projectors that can still be used is the hand-cranked Pathé 9.5mm designed for the home market. Although not really old—many of them date from the 1920s—they are decidedly workable, and, an important point, 9.5mm film can still be found in junk and second-hand shops. One must be truly enthusiastic to attempt to refurbish for use aged cine cameras, though devotees with the requisite photographic knowledge can, with some effort, make their own film stock. One point that should be born in mind is that the early film made from cellulose nitrate is highly inflammable and unstable, and has long been replaced by the safe cellulose acetate.

It is impossible to say how much early cinematograph equipment exists today outside science museums, in comparison with the considerable quantity of period cameras. It must be emphasised that not all plate cameras are necessarily old; they are still widely used by professional photographers. Likewise, it is difficult to say how much early film still exists. Because of the high fire risk involved it is probable that much has been deliberately destroyed, though it has been the policy of film studios to stock-pile antique film. Although this has had the effect of removing it from the open market, a certain amount of invaluable material has been preserved, and, transferred to modern film, is a welcome sight on television screens.

The various optical toys, the fantascopes, the stroboscopes, and zoetropes, should be snapped up. They share with magic lanterns a desirable quaintness, and prices are already reaching a realistic level.

CHAPTER 7

AUTOMATA AND MECHANICAL TOYS

To ASSUAGE the wrath of collectors of automata it might be wise to give a 1586 definition of a toy: 'A plaything for children or others; also, something contrived for amusement rather than for practical use.' The makers of automata were the Mozarts of the mechanical world, and it is impossible to over-estimate the ingenuity and delicacy of their work. It hardly needs mentioning that for many years automata have been highly regarded, though unquestionably prices have not yet reached their ceiling. It is a sphere for those with bulging wallets.

Prior to the eighteenth century there were two areas in which precision work was carried out—firearms and watch and clock-making—and the makers of automata have much in common with the watchmakers, in particular their ability to work on a small scale and their manipulation of clockwork. Most automata depend on the driving power of the spring, though there are automata powered by water, sand, compressed air, steam, and even mercury. The true automaton, once activated, does everything without further human interference, and so differs from pseudo-automata in which an operator guides or controls.

An automaton is a machine acting by means of a concealed mechanism. Early simple examples are to be found in accessories surrounding clocks, such as striking jacks, or revolving skeletons, activated in a straightforward manner by the clock mechanism. Such automata are considerably less sophisticated than those constructed by Hero of Alexandria (*fl.* about 100 BC) powered by water and steam, and the tradition of hydraulic automata was preserved by the Arabs, leading up to the large number of hydraulic automata found in gardens and grottoes of the sixteenth and seventeenth century, technically no more adventurous but installed with a profusion of detail. The principle was simple; water was channelled so that its weight set into motion what a sceptic of 1644 called 'all this gimcrack ironmongery'.

The striking jacks of clocks—the oldest one in the world being in Wells Cathedral and dating from 1392—have a direct relevance to the clock itself. A sequence of figures moving around on the hour or half hour, as exemplified in the modern town clock in Coventry or the superb Victorian example in the old Birmingham Market Hall destroyed during the war, have rather less, and in picture-clocks the time is of less interest than things happening. The picture could take the form of a ship moving up and down on waves, a locomotive coming out of a tunnel, or a balloon ascending, or all three. Picture-clocks have a respectable ancestry, dating back to the end of the sixteenth century, but most of those seen today are relatively modern. They are one of the few kinds of automata within the price range of the average collector, but the later ones, with components made of cardboard, are usually grotesquely overpriced and have no relation to the earlier ones. The category picture-clocks merges into the mechanical picture, immensely intricate and made with the precision of a watch. An eighteenth-century mechanical picture at the Paris Conservatoire des Arts et Métiers includes in its programme a laundress washing linen, a peasant girl beating butter in a churn, two fighting cocks, a fisherman raising and lowering a rod, a drinking scene with soldiers, a carriage crossing a bridge, a church tower with a bell that rings, and clouds drifting across the sky, occasionally allowing a ray of sunshine to illuminate the scene. There are no less than fifteen individual movements, activated by a strong spring connected to wooden cylinder, in which are set pins and staples which engage levers with rods. Other mechanical pictures are operated by rods and spokes geared and connected to a spring. The spring also drives one or more endless chains for processions of soldiers, farmyard animals, peasants etc. On some pictures the spring also puts into motion a pendulum backing on to subjects such as swings.

These mid-eighteenth-century marvels are surely among the most advanced products of their day, but ingenuity could go even further, as in the miniaturisation necessary for the construction of scenes in watches or snuff-boxes, or singing-birds incorporating a tiny bellows to produce the bird-song via a whistle, the pitch of which is altered by means of a piston sliding in and out. The ability to work on a minute scale was not confined only to watch-makers. As early as 1578 a blacksmith made a 48 link chain of gold small enough to put round the neck of a flea, and in 1745 an exhibition in London of 'miracles of art' included an ivory open chaise with the figure of a man in it drawn by a flea. No less remarkable was a pair of steel scissors 'so small that six pair may be wrapped up in the wings of a fly'.

The fusee is a cone with a spiral groove, attached to the side of the first wheel of the watch, and connected with the barrel or drum containing the main-spring by a chain, hooked, at its ends, to both. The figure to the right, in the above cut, is the fusee; that to the left is the drum.

62 *The fusee movement, much used in superior watches and automata.*

We are therefore inclined to be over-impressed by the fine work carried out in the small confines of snuff-boxes, watches, and imitation birds, forgetting that the makers were no less able than are those of today; workers on the smallest of scale are not dependent on technical advances. It must also be remembered that the makers of automata were masters in the exploitation of clockwork; the triumph of the spring dated from the sixteenth century with the invention of the fusee, where the mainspring turns a barrel on which is wound a piece of catgut (later a chain), the other end of which is wound upon a spiral drum, so that as the spring runs down the tension remains. By the end of the eight-eenth century, tiny units worked by springs had reached a height of perfection that it was impossible to supersede, and watches containing automata of the period work as beautifully today as they did then.

In the use of what might be termed small power-packs, the eighteenth and early nineteenth century were notably deficient, relying exclusively on the power of the spring, and it is there-fore not surprising that this mastery was shown off not only in

63   *A very fine French automaton, the Dandy.*

small objects such as rings and watches, but in larger items, such as mechanical animals, 'moving anatomies', and, of course, mechanical pictures.

No account of automata would be complete without mention of a handful of masterpieces. One of the most impressive is the peacock in The Hermitage, Leningrad, which first appeared in an inventory of 1811 and is believed to be the work of an Englishman, James Cox. Accompanied by a cock, a squirrel and an owl, the peacock perches on a gilt bronze tree. The tableau is nine feet high. At the side of the tree are gourds and mushrooms. The top of one of these has two small openings in it, and in these can be seen the hours and the minutes, and at one time a cricket on top of the mushrooms used to jump up and down sixty times a minute, acting as second hand, for this marvellous creation was also a clock. The peacock shakes its head, opens out its tail, and rotates in a circle, the cock flutters about and sings, and the owl moves its head, opening and closing its eyes.

Yet even more famous is the automaton duck made by Jacques de Vaucanson and first exhibited in 1738, and described at the time as 'an artificial duck made of gilded copper who drinks, eats, splashes about on the water, and digests his food like a living duck'. Enormously complex, with thousands of working parts, the duck was a wonder of its age, but as the years went by it became more and more neglected, until when Goethe went to see it in 1805 it was almost completely paralysed. 'The duck had lost its feathers and, reduced to a skeleton, would still bravely eat its oats but could no longer digest them.' In an age of talented watchmakers there was still no one with the expertise to repair and renovate the duck, for not only were there mechanical problems. The duck ate, digested, and *excreted* the grain, and how this was managed puzzled the cognoscenti. Eventually Rechsteiner, a celebrated clockmaker, was found to undertake the task, which took him three and a half years. The duck was on show in Paris in 1846, but when it went wrong again Rechsteiner was making a duck of his own, in which all the features of the earlier duck reappeared, including the ability to defecate, noted by an 1847 reporter: 'The truth is that the smell which now spreads through the room becomes almost unbearable.' The number and variety of movements necessary would have meant a spring of mammoth proportions, and Vaucanson used instead a weight

64 *A ballerina who dances in a realistic manner and turns her head.*

hung by a cord wrapped around a drum. Thirty different levers directed the movements, while the duck itself contained its own complicated mechanism, the most astonishing being the means of digestion, which seem to have been chemical, and closely related to that of a real duck.

Even more astounding are the writing and drawing automata. The first were made by Friedrich von Knaus between 1753 and 1760; there were four in all, the first three consisting of a hand holding a pen atop a metal globe, the last a small figure holding a pen. The mechanism was in the globe, not the hand or figure. Von Knaus's automaton could write as many as 107 words. The ingenuity of the mechanics needed for this feat makes one wonder why relatively simple machines such as the typewriter were not invented for another 120 years.

The writer invented by Pierre Jaquet-Droz was even more sensational, a 28in replica of a small boy. When the mechanism contained in the wooden body of the boy was started, the boy dipped his pen into an inkwell, shook it twice, put his hand at the top of the page, and then paused. Pressure on a lever set the automaton working. The letters were formed carefully, using both thin and thick strokes, with gaps between the words, and at the end of the line the hand would move lower down the paper to commence a new one. Two fusee movements drove the automaton, and the writings were programmed by a disc with wedges at the bottom of the mechanism. The disc could be taken out and the wedges adjusted to change the writing matter. The apparatus was so sensitive that a change of temperature produced a scrawl rather than a coherent message, and certain moving parts were faced with rubies to avoid the slightest friction.

Henri-Louis Jaquet-Droz, the son of the maker of the writer, created a drawing automaton known as 'the Draughtsman', also driven by two fusee movements, which although mechanically simpler created even more of a stir, executing four Baroque-style drawings. The Jaquet-Droz family were also prominent in the field of snuff-boxes and other small pieces, and made a pair of engaging musical automata, a lady playing an organ, operated by a clockwise-driven pinned barrel, and a lady playing a harpsichord, the first playing five melodies, the second either sixteen or eighteen. A writer in *Chambers's Journal* in 1876 stated that

65   The Indian conjuror with the mechanism exposed.

Jaquet-Droz (whom he calls Le Droz) was responsible for an
automaton peacock, and it is within the realms of possibility, considering the lack of reliable data, that the Leningrad peacock
may be the work of the Jaquet-Droz family. It is certainly a
more likely source than an obscure Englishman; the English have
rarely shone in the manufacture of automata, and during the
period in which the Jaquet-Droz family flourished the Swiss and
the French completely dominated the watch market, and the
associated automata.

It is a far cry from such splendiferous creations as the writer
and the draughtsmen to the automata animals, the smaller specimens of which come between the categories of costume jewellery
and toys. Articulated caterpillars made of gold, enamel and
pearls, gold and pearl-encrusted mice with eyes made of rubies,
rabbits coming out of lettuces, automata lizards with a very life-
like tail movement—it is difficult not to feel that this is ingenuity
run mad, gewgaws for an effete and jaded society.

The line between automata and toys is truly crossed with such
products as the swimming doll exhibited at the Exposition Uni-

verselle of Paris in 1879, and the turn of the century saw the
Martin family of France marketing six inch automata, with titles
such as 'the messenger boy', 'the little pianist' and 'the barber'.
By that time the world was not so much in thrall to clockwork,
and forces such as electromagnetism were used to act externally
on figures. Such figures naturally lose their status as automata,
and can be classed with simple toys of the kind in which figures
in a glass box rush about madly when the glass is rubbed with a
soft cloth (creating electricity), or the magnetic toys popular
amongst Victorian children, in which cardboard fish with a metal
tag are drawn from a box by rod, line, and magnet.

The peacock, the duck, the writers, draughtsmen and musical
performers, these serve to fill in the picture but are hardly likely
to be met with except in museums (the Jaquet-Droz pieces are in
a museum in their home town Neuchatel, the duck is lost, but
England has an automaton swan in County Durham). They are
interesting in that they show how advanced clockwork technology was, and how watchmakers and engineers could create
from virtually nothing a refined transmission system that has
never been surpassed. It is also significant that England played
only a minor part in this field, and one wonders what would
have happened had the Industrial Revolution been sparked off in
Europe rather than in Great Britain. What would have occurred
had the talents of brilliant men like Vaucanson been directed into
areas other than trivia to amuse the rich?

The true combination of automaton and toy is the walking
doll. The simulation of human movement was a challenge, and
although the movements of arms and heads, and even fingers,
could be reproduced faithfully, as could the action of legs, the
problem of balance was only solved by having the centre of
gravity very low, and hidden by long skirts, or by supporting the
figure. A walking doll pushing a perambulator was a way of
getting over the difficulty, and so too was the provision of reins.
Leg movements were avoided completely by some makers by
using wheels, which were hidden by the long skirts. As dresses
of the period were full-length, skirts touching the floor were by
no means anachronistic. Even the better types of walking automata avoided the question of the jointed leg. One of the cleverest
walking dolls, the Autoperipatetikos, invented in America and
patented in England in 1862, had one-piece legs. A bar extended

66   A drinking bear, more of a toy than an automaton.

down the leg, and through a hole in the sole of the foot, raised the leg as the other leg swung forward. The bar on this leg extended as the first leg completed its follow through, and so on. The impression given was of a sliding movement combined with walking on stilts. The term legs is perhaps a misnomer, for the circular base of the mechanism was set very low, just above the ankles. A number of Autoperipatetikoses are about, varying in costume and external detail; some have china heads, others heads of bisque, papier mâché, or wax. Most of these dolls were female, but one at least was male, clad in a Zouave costume.

In 1868 William Goodwin of New York patented an automaton doll with jointed knees, but even celebrated makers of automata who had turned to the manufacture of dolls because of the high demand preferred simpler methods of propulsion, such as wheels. A true walking doll patented in Paris in 1890 by a clockmaker, Jules Steiner, was intended to imitate a baby's first steps, but the *Bébé Premier Pas* had to be held by the shoulder or the hand.

Even more stimulating to inventors than the walking doll was the talking doll, and as early as 1778 Baron von Kempelen produced a kind of talking doll, followed in 1780 by Kratzenstein. One of the first uses of the phonograph was its insertion in the body of a doll. An incentive was given by the award of a prize by the Imperial Academy of Science at St Petersburg for the reproduction of vowel sounds. The prize was won by von Kempelen, who devised a hollow oval box divided into two parts, and fitted with a hinge. The hinged box caught sound coming from a tube fitted to a bellows, and by opening and shutting the 'jaws' *a*, *o*, and *u* could be produced, plus an imperfect *e*, though *i* was impossible. After several years experiment, von Kempelen also managed to evoke *p*, *m*, *and l*, but the effect was still imperfect, and he went on to construct a mouth made of rubber allied to two tin tubes, and this object could be made to enunciate quite complicated words, such as *opera* and *astronomy*. Goethe saw the device, and said that it was 'able to say some childish words very nicely', though von Kempelen remained unsatisfied, conscious that he had failed, and that his listeners were frequently deceiving themselves; *d*, *g*, *k* and *t* were merely modifications of *p*.

The problem appealed to English inventors, and Robert

Willis of Cambridge produced an improved version of the von Kempelen apparatus. As yet it was a purely scientific matter, with no thought of inserting the mechanism into a doll, though this was soon remedied, and in 1815 another Englishman, Robertson, made a waxwork doll that was said to be able to pronounce all the letters of the alphabet. Robertson had been to Russia, where encouragement was still offered to inventors of talking machines.

MILLIKIN AND LAWLEY'S
WORKING MODEL PADDLE

STEAM BOATS, 5s. 6d., 7s. 6d., 10s. 6d., 15s. 6d., £1. 1s. SAILING YACHTS, 5s. 6d., 7s. 6d., 12s. 6d., 15s. 6d., £1. 1s., £1. 10s., £1. 15s., £2. 2s. MODEL STEAM ENGINES, 5s. 6d., 7s. 6d., 10s. 6d., 15s. 6d., £1. 1s. to £10, 10s. Beautifully Illustrated Catalogue of Model Engines, Marines, Locomotives, Separate Parts of Engines, &c., Dancing Niggers, Schooners, and Yachts, fully rigged, post free, 1s. 3d.

CLOCKWORK PADDLE STEAM BOAT, £1. 1s. Works for a long time in water.

CLOCKWORK SCULLER, 10s. 6d.— This marvellous automaton rows his boat in a most life-like manner, in water, affording rare amusement.

CLOCKWORK JUBA DANCERS,

12s. 6d., post free 13s. 6d.; single figure, 7s. 6d., by post 8s. 6d. Very amusing and marvellous-ly life-like automaton dancers. These clockwork dancers afford a great attraction at fancy fairs, fêtes, bazaars, &c. Illustrations of various clockwork figures in "Our Magazine," 15 stamps.

COLLEC.TIONS of PRETTY NOVELTIES for SALE at FANCY FAIRS, BAZAARS, &c., comprising very saleable and useful KNICKNACKS, 7s. 6d., 10s. 6d., 15s. 6d., £1. 1s., £1. 10s., £1. 15s., £2. 2s., £2. 10s., £3. 3s., £3. 10s., £4. 4s., £4. 10s., £5. 5s.,

67 A selection of mechanical toys on sale in 1875.

A talking automaton by Professor Faber was bought by Barnum the circus proprietor in 1853. Although the bellows mechanism and tubes follow previous automata. Faber used a mobile tongue made of ivory in the sound-producing mechanism and pieces that served as teeth and lips, and the Faber talking automaton was unquestionably the most sophisticated made up to the mid-nineteenth century. The various components were brought into use by a kind of piano keyboard. Faber even had a tiny mill to produce a rolling r, and for nasal sounds a separate cavity could be connected up by means of a lever. Although more artistically served up, the talking doll with which Johannes Mälzel won an award in 1823, perhaps the first authentic speaking doll rather than mechanism, was of less consequence. A bellows alternately pumped by the right and left arm forced air through a reed to the voice box, and an explosive double sound 'Pa-Pa' was emitted. Muffled by a small gadget, 'Pa-Pa' became 'Ma-Ma'.

All nineteenth-century dolls, especially those with faces of wax or china, are eagerly collected, and talking dolls more so than most. When there were many to make the exterior of dolls but few to provide a mechanical interior it is not surprising that automata of all kinds are usually finely finished.

The mice and lizards made from gold, pearls, and enamel are light years away from common or garden mechanical toys using the same principle, clockwork. The expansion of the tin-plate industry created the perfect material for making toys with cheap clockwork movements. Between 1834 and 1860 British tin-plate production multiplied ten times; between 1860 and 1870 it doubled. The construction of tin toys and the fitting of movements was suited to unskilled labour. Germany soon dominated the toy market, Nuremburg and Fürth, the latter specialising in tin toys, being particularly productive.

It is very easy to rhapsodise on the subject of automata, but this is what we are meant to do. We are the willing victims of eighteenth-century confidence tricksters, our admiration of the mechanics encouraging us to overlook the fact that as objects automata are ostentatious and vulgar. Rare among the items that come within the category of mechanical antiques, they are also useless. There is some social excuse for the making of dolls and mechanical toys for children, but the expertise and fantastic sums of money that went into creating such things for adults are ludicrously inapposite.

There is also an inbuilt tendency to overrate the abilities of the automaton makers. At the simplest level it is surely naïve to

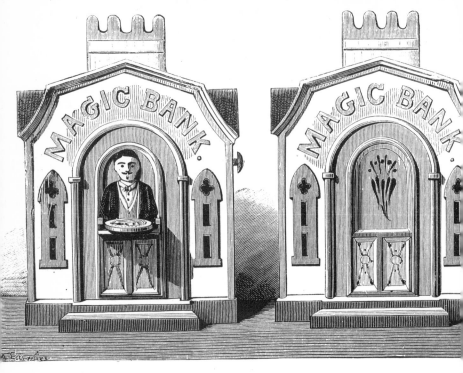

68 *Toy money boxes were the poor relations of automata. This one worked on a simple spring mechanism.*

enthuse over handmade mechanism because it is handmade, and although automaton makers may be the Mozarts of engineering, this analogy could not be pursued in terms of creative ability or imagination. Nor did the automaton makers 'leap into the dark' as nineteenth-century inventors did when confronted with the mysteries of electricity or acoustics. Like their products, they were wound up and the gearing and levers did the rest.

Even the most ordinary automaton, accompanied by the most routine of musical movements, will never make less than a hundred pounds. A nodding head, an arm moving jerkily in the perfunctory accomplishment of some task, a lip-moving and eye-rolling head that could today be made by any sixth former with a mechanical bent, these are treated with awe if the objects have any kind of pedigree at all.

Eighteenth-century automata are naturally invested with far more kudos than those of the nineteenth century, it being erroneously accepted that the eighteenth century was populated by idiots with the exception of a few furniture-makers, architects, watch and automata makers, and the odd writer, painter and musician. The automata trade dwindled in the nineteenth century because patronage shifted; the expensive and useless were still to be found, but in other fields.

The aura surrounding eighteenth-century automata has extended to their makers, Vaucanson, the unknown Robertson, Rechsteiner, Jaquet-Droz, and von Kempelen, and their renown has eclipsed later automata makers who uncompromisingly made automata to earn money, such as the conjurer Robert Houdin (1805–71), who made pipe-smokers, acrobats who performed tricks, and orange-trees that appeared to grow fruit under the eyes of a paying audience, or Professor Pepper (1821–90), master of optical tricks and originator of 'Pepper's Ghost' whose automaton trapezist fascinated the groundlings.

Pseudo-automata are robbed of their appeal when it is found that there is a human element involved, and child-like wonder is thwarted. Notwithstanding the intricate machinery involved in the construction of von Kempelen's automaton chess-player, interest dropped considerably when it was discovered that the automaton's prowess at chess was not magical, but the result of a man concealed in the bowels of the machine. The inventor encouraged the deception, and does not seem to have had any ethical problems when the preface to a pamphlet on this pseudo-automaton stated :

The most daring idea that a mechanician has ever ventured to conceive was that of a machine which would imitate, in some way more than the face and movement, the master work of Creation. Von Kempelen has not only had the idea, but he has carried it out and his chess-player is, indisputably, the most astonishing automaton that has ever existed.

The fashionable world goggled at the chess-player, believing in it because they wanted to believe in it. It became the most famous automaton in the world, and was taken all over the world along with its hidden operator. Very few were not taken in by it, and Edgar Allan Poe was by no means popular when he wrote an article proving with incontrovertible logic that the chess-player could not be a machine.

The sensation caused by the writing automata also encouraged wily inventors to produce pseudo-automata. Eighteenth- and early nineteenth-century high life was neurotically fond of being blinded by science, and was as gullible as the music hall audiences of the latter part of the nineteenth century when faced by the pseudo-automaton Psycho, joint invention of the famous conjurer J. N. Maskelyne and John Clarke. Psycho, presented at the Egyptian Hall in 1875, played whist with the audience, did arithmetical problems, and answered questions by striking a bell. Isolated on a hollow cylinder of transparent glass, Psycho appealed only marginally to the fashionable world, though had it been exhibited in the Court of Vienna or St Petersburg instead of to the *hoi polloi* it would probably have received the same degree of bemused adoration as the chess-player and its contemporaries.

CHAPTER 8

MECHANICAL MUSIC

ALTHOUGH MECHANICAL music reached its zenith in the nineteenth century, the performance of previously programmed music goes back many centuries. The inventory of the Duke of Modena's collection of organs and other keyboard instruments drawn up in 1598 contains references to the *Organo Tedesco*, or barrel-organ. The barrel-organ was known by a variety of names throughout Europe; in England it was also known as the grinder-organ, the street-organ, and the hand-organ, in France as the *orgue d'Allemagne, orgue mécanique*, and *cabinet d'orgue*, in Germany as the *Drehorgel, and Leierkasten*, and in Italy as the *organo tedesco* and *organetto a monovella*.

The barrel-organ owes its name to the cylinder, set with pins or staples, which when it revolves activates the valves of organ pipes, which are thus mechanically opened, admitting air from a wind-chest. Each revolution of the barrel plays one tune, but a notch-pin in the barrel head, furnished with as many notches as there are tunes, enables the performer to shift the barrel sideways and change the tune. A system of simple gearing keeps the cylinder turning at a uniform rate.

Barrel-organs come in all shapes and sizes, and although normally powered by hand, a large stationary barrel-organ worked by hydraulic power was illustrated in a book of 1615. The earliest barrel-organs were probably made in the mid-fifteenth century in the Netherlands (thus one of their English names, the Dutch organ). Many were portable and primitive, with a compass of only twenty-four notes; others were sufficiently elaborate for

69   *Although barrel pianos were mostly used outdoors, parlour models were also used,*
echoed by this cartoon of 1895.

MISS PINCKNEY : "Won't you play something on my new
Christmas present, Professor ?"
PROFESSOR MECHBACH : "Ohf it vas your bleasure."

PROFESSOR MECHBACH : "Oxcuse me, der piano vas locked."
MISS PINCKNEY : "How stupid of me ! It's one of those dear,
delightful Italian instruments, you know.   You'll find the handle

Mozart to write for them. In 1737 Horace Walpole, the collector and dilletante, wrote: 'I am now in pursuit of getting the finest piece of music that ever was heard; it is a thing that will play eight tunes. Handel and all the great musicians say that it is beyond anything they can do, and this may be performed by the most ignorant person, and when you are weary of those eight tunes, you may have them changed for any other that you like.' This particular organ was put in a lottery, and made £1000.

It need hardly be said that instruments of this status are quite out of reach of the average collector, and much the same is true of the hurdy-gurdy, which is of even earlier pedigree, dating back to the thirteenth century at least. The hurdy-gurdy is generally shaped like a guitar, and consists of a sound box, enclosing a wheel covered with rosined leather, worked by means of a crank at the tail-end of the instrument. The hurdy-gurdy had a short keyboard, and four or six strings. When a key was depressed, it would 'stop' a string (or all the strings in earlier versions); when one string was stopped, and when the wheel was turned, the other strings would respond in sympathy.

The effect was of a drone, somehow similar to bagpipes, and although the hurdy-gurdy was hardly sophisticated, it retained its appeal until the eighteenth century, especially in France, and stimulated such musical curios as the *Geigen Clavicymbel* (c 1600), a harpsichord in which the strings were set in vibration by a parchment-covered wheel set in motion by treadles.

The barrel-organ is often confused with the barrel-piano, once a common object on the London streets. The same principle is used—a revolving barrel set with pins, which activate tuned wires instead of wind pipes. The barrel-pianos brought music to the poorest people in the towns and cities, and although loud and strident, and except to enthusiasts terribly jangly, the music they perform recaptures the spirit of the age. They provided a living for clannish expatriate Italians and especially Italian children who hired the instruments with or without monkeys from a centre in Clerkenwell.

The principle of the revolving barrel set with pins was also used with the most characteristic providers of nineteenth-century mechanical music, the cylinder musical box, invented in 1796 by Antoine Favre of Geneva. The same principle was used in certain

70 *A fine early musical box by the famous makers Nicole Freres.*

kinds of automata, such as mechanical pictures. The power source for automata and musical boxes was also the same—clockwork. The musical box, unlike automata, barrel-organs, or barrel-pianos, was direct acting; the revolving cylinder acted on tuned teeth on a comb, the tips of which were set vibrating by the pins of the cylinder.

From being a tinkling gimmick for incorporation in watches, snuff-boxes, or automata, musical boxes became larger and valid musical instruments. From about 1815 a number of watchmakers gathered in Switzerland, creating finer and more sophisticated musical boxes. By staggering the pins, as in some barrel-organs, a large number of tunes (as many as twenty) could be put on one cylinder, and there were some musical boxes which used interchangeable cylinders, though these were prohibitively expensive except to the very rich.

The snag about musical boxes was that no matter how exquisite the material and cunning the arrangements of operatic arias and overtures the tone was very much of a muchness, etherial, twinkling, but lacking in variety. Refinements only served to disguise this; a zither attachment used tissue paper, repeated notes emulated a mandoline, and drums and bells, tuned to the teeth, added an extra. Chordal arrangements of hymn tunes were also very popular, but chords highlighted the major deficiency of the musical box—the inability to hold a long note. A trill instead of a sustained note did not solve this.

More successful in the production of sustained notes and chords was the mechanical organ operated not by a pinned barrel but by a perforated paper roll. Where there was a hole in the paper sound was produced. Such mechanical organs could be small, such as the Organette of 1878, or large, such as the Orchestrion of 1887, and could be augmented with drums and other percussion. The small organs were basically an American innovation, and were a valid substitute for a harmonium where there were no performers available, and although they can be found in some quantity, their tone soon gets tiresome, and they lack the charm, and the intricate precision-made mechanism of the musical box.

These organettes, sold for a pound or two, had one advantage over the cylinder musical box. In theory there was no limit to their repertoire, which was only restricted by the quantity of rolls

71   *An interchangeable cylinder musical box.*

one could afford to buy. The perforated roll was to provide a stimulus to inventors when the time came for the player piano.

The concept of interchangeable parts that we have met in the chapters on the sewing machine and the typewriter had no place in the philosophy of Swiss watchmakers. Each musical box was individually made, and it is a credit to the workmanship involved that these boxes are functioning today as well as they ever did without having anything done to them in a hundred years. Even in the 1870s, when production was stepped up to provide gaudy inferior boxes for the rising middle classes, no answer had been found to the problem of the small repertoire. Few musical boxes carried more than ten tunes, and repeated performances could prove, to say the least, wearisome.

In 1886, Paul Lochmann of Germany designed a musical box that would play a disc instead of a cylinder. As the disc revolved on a spindle, notches operated on tuned teeth in the same way as the cylinder musical box. The cylinder box was small in size and tone. The disc musical boxes, known generally as Polyphons (the principal maker) were large and loud; some were ten feet tall. The use of cardboard discs was soon discarded as they wore out quickly, and metal discs were adopted. In 1889 two German firms, Symphonion and Polyphon, began to make disc musical boxes, as did the Regina firm of America.

Robust and indestructible, the German machines were imported into Britain in quantity, and although hoteliers and innkeepers recovered capital outlay by having the machines fitted with a penny-in-the-slot mechanism, they were amusements for the wealthy, status symbols, expensive furniture that played a tune. Discs could be cut in response to the success of a drawing-room ballad or music hall tune, and it seemed as though the disc musical box was here to stay. Its vogue, however, was a good deal shorter than that of the musical box, and by the outbreak of the first world war it was dead, killed by an unconsidered novelty, the phonograph and its successors.

Whereas the story of the musical box is straightforward and uncomplicated, that of the phonograph is a muddle. It followed closely on the heels of the telephone, and was considered even by its inventors to be an aid to business rather than an instrument of entertainment.

The men most deeply involved in the evolution of the 'talking

72 *A polyphon disc musical box.*

machine' had cut their teeth in telephony and telegraphy. In 877 Thomas Edison was working on an instrument that transcribed telegrams by indenting a paper tape with the dots and dashes of the Morse code, and later repeated the message any number of times or at any speed. He observed that when the paper tape ran through at high speed, it struck a spring that gave off 'a light musical rhythmic sound resembling human talk heard indistinctly'. He decided to experiment further, and by using a diaphragm with an embossing point he succeeded in imprinting speaking vibrations in a moving strip of paraffin paper.

Edison dropped the idea of telegraphic improvement in favour of a talking machine. Mechanical music was in thrall to the principle of the revolving cylinder, and instinctively Edison used this principle. The first phonograph consisted of a grooved metal cylinder wrapped around with tin foil, two diaphragm-and-needle units and a spindle through the cylinder with a handle on the end. The recording needle followed the spiral grooving and the spoken vibrations created a 'hill and dale' pattern in the trough in the groove. The reproducing needle converted these patterns into sound. The first words spoken into a phonograph were 'Mary had a little lamb' and these were played back to an astounded Edison.

In an America that thrived on patent troubles, Edison initially had a clear field, seemingly the only man venturing into voice reproduction, but in France Charles Cros was experimenting with tracing sound waves on lampblacked glass and photo-engraving these tracings. Cros did not have the enterprise of Edison; in December 1877 Edison went to the offices of the *Scientific American*, which once again proved a valuable ally to native inventors. The phonograph was well worthy of a paragraph :

> Mr. Thomas A. Edison recently came into this office, placed a little machine on our desk, turned a crank, and the machine inquired as to our health, asked how we liked the phonograph, informed us that it was very well, and bid us a cordial good night. These remarks were not only perfectly audible to ourselves, but to a dozen or more persons gathered around.

The craze was soon over. The tin foil phonograph was im-

73   *The mechanism of a disc musical box, illustrating nineteenth century mechanical know-how at its best.*

erfect, the words were croaks, the mechanism was crude, and
espite the Parlor Speaking Phonograph for home amusement
$10, capacity 150–200 words) prospects were poor. Tin foil
vore out after half a dozen performances; total playing time—
ess than ten minutes. A dollar a minute was expensive for a toy.
V. H. Preece, who had brought the telephone to England in
877, had a boyish interest in all novelties from the United
tates. His pronouncement on the phonograph was the under-
atement of the year: 'The instrument has not quite reached
hat perfection when the tones of a Patti can be faithfully re-
eated . . .' five hundred tin foil phonographs were produced
efore the fad went cold, and Edison was persuaded into a more
nmediately encouraging field, the invention of the incandescent
amp.

But fresh from the triumph of the telephone came Graham
Bell, and with his cousin Chichester A. Bell, a chemical engineer,
nd Charles Tainter, an instrument-maker, he decided to carry
on where Edison had left off. They first of all discarded tin foil,
ubstituting wax, and in place of the rigid needle they adopted a
oosely fixed stylus. They were granted a patent in 1886, and in
889 the graphophone was put on display. It was a great
dvance on the Edison apparatus; the wax allowed closer groov-
ng, and the primitive direct drive of Edison's phonograph gave
vay to a foot-treadle mechanism or an electric motor powered,
of course, by cumbersome batteries. Edison was incensed by this
ntrusion into his field, dropped his electrical work, and produced
a revitalised phonograph. He made only one advance on the
raphophone; the graphophone had a wax-coated cardboard
ylinder, Edison's machine used one of solid wax.

The graphophone was designed as a business auxiliary, and in
888 it began to make commercial headway. In the same year
Edison had a shot at recording serious music, and the pianist
Hoffmann consented to perform; so did Hans von Bülow, who
played a Chopin mazurka, one of the few self-contained pieces
hat could be slotted into two minutes.

The patent wars that had dogged the progress of the sewing
machine and the typewriter now started around the talking
machine, but it was not Edison who sued Tainter and Bell (they
had artfully patented a process of 'engraving', he had specified
embossing or indenting') but the other way about. The inter-

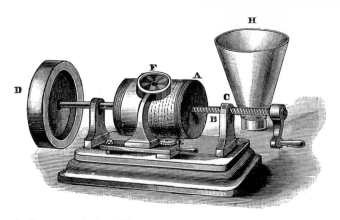

74 A diagram of the Edison phonograph.

vention of big business halted the war, and the North American Phonograph Company encompassed both the aggrieved parties. The fate of the phonographs and the graphophones was now even more closely bound to industry.

But with big combines there are always subsidiary companies who want to make money on their own. The talking machines were expensive, so, as in the case of the disc musical boxes capital outlay was offset by penny-in-the-slot mechanisms. Penny-in-the-slot phonographs were therefore exactly contemporary with penny-in-the-slot disc musical boxes; in value for money, there is no question which won—the disc musical box.

The disc musical box had the advantage over the phonograph that it had had over the cylinder musical box; the cylinders of both the latter were one-offs. Each cylinder was unique. Bands worked overtime to record marches and similar music that would not call attention to the defects of the recording instrument. By 1893 one large company, Columbia, had a thirty-two-page catalogue; one band alone had recorded eighty-two marches. But the Edison electric phonograph was still $190.

It is interesting to observe that where electric motors were needed, as in the vacuum cleaner, they were not used, whereas the phonograph, which did not need it, was made prohibitively expensive by being equipped with one. It is clear that Edison did not bring to the evolution of this particular brainchild the common sense that was his wont. He had persisted in the inept use of tin foil, and he had not taken notice of the motive forces of successful mechanical music-makers—clockwork. This was remedied by one of his employees. In 1894 the clockwork-powered Graphophone Grand was put on sale for $75.

Like its competitor in music for the masses via a coin-in-the-slot, the disc musical box, the phonograph was doomed even in success, for in 1887 Emile Berliner, a thirty-seven year old German-American, had shown the way to the future. Berliner had been in telephone research, but was sufficiently interested in the phonautograph of 1857, a laboratory instrument used in measuring and analysing sound by transcribing vibrations on to lampblacked paper, to venture into talking machines. Instead of a cylinder, Berliner chose a disc; he called his machine a gramophone.

Berliner made a false start (as Edison had done) by using coated zinc, and, a process common in engraving, acid to bite into those parts of the zinc he wanted etched out, while the coat was impervious. He was in the van in two ways—copies could be made of the matrix, and the gramophone was designed for home entertainment. His most important innovation was that, as the disc rotated, the recording needle cut the wax not from the bottom of the groove but from the sides.

Berliner did not persevere with the gramophone with the diligence it deserved. It was served up as a toy in Germany, the makers of which obtained a licence from him, and produced miniature hand-propelled gramophones with five-inch celluloid or rubber discs which were sold all over Europe and are occasionally met with today, commanding slightly higher prices than they did when first produced in 1889 (two guineas including six records).

In 1893 Berliner thought the time was right for exploitation in America. He had not the prestige of Edison, and backing was difficult to obtain. In 1894 the first gramophone 'plates' were issued, and three different types of gramophone were marketed, the most popular of which was the Seven-Inch Hand Gramophone. Like the phonograph-makers, the United States Gramophone Company neglected the obvious method of power, clockwork, and the first gramophone had a direct drive, which made a consistent speed impossible to obtain. Berliner did not make the mistake of overpricing his merchandise. The Seven-

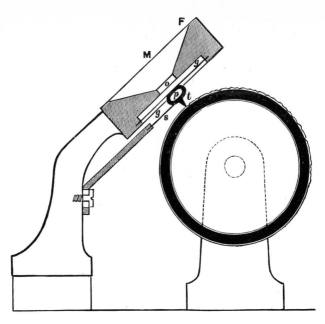

75 *Diagrammatic section of the phonograph.*

Inch Hand Gramophone sold for $12, but even at this price there were few buyers, and it was not until the following year that a syndicate was formed to promote the gramophone, which was decidedly the poor relation.

One disadvantage of the phonograph had been its lack of volume, but a horn was now provided instead of rubber hearing tubes. By the turn of the century the phonograph was priced down to $7.50, and the cylinders could now be duplicated. A master cylinder would produce twenty-five copies before it wore out but production did not match demand. In 1901 a successful process was introduced to mould cylinders, rather than to use one master to trace the grooves on to another, but this was shutting the stable door after the horse had gone, for there was no overlooking the fact that the gramophone would eventually replace the phonograph, though, surprisingly, the phonograph survived until 1929. In view of the high price that phonographs get today, this little-known fact should be borne in mind, and a close inspection of a likely buy should be carried out. £50 or more might well seem a high price to pay for a product of the late 1920s.

1896 was a key year for the gramophone : a satisfactory clock-work motor was evolved, and for the first time it was judiciously promoted, with the slogan 'The Talking Machine That Talks Talk !' The Improved Gramophone of 1897 was the first gramo-phone that still exists today in any quantity. 1897 saw the introduction of shellac discs instead of vulcanised rubber, and a trade war between cylinder and disc led to a vicious patent war, resulting in a schism between the various parties involved in the production of the gramophone. A breakaway gramophone group invented the zonophone, in which the clockwork was enclosed in the case instead of situated at the back. By the turn of the century the Improved Gramophone was down to $25, within reach of all, and the discographers of the future could buy a hand-driven model for $3.

These dramatic events in America had, as one might have expected in view of the tardy arrival of earlier American in-novations, left Europe cold. Despite the enthusiasm with which Britain had greeted the penny-in-the-slot disc musical box, the penny-in-the-slot phonograph had not caught on at all, though in France the Pathé Brothers, busily involved in the youthful cinematographic business, saw the possibilities, and produced a cheap cylinder machine 'Le Coq'. Britain was never really involved in phonographs, though it fell to shock tactics when the Gramophone Company began operations in 1898, using discs stamped in Germany, and within two years the company was advertising 5000 different records. Prior to 1900 England was forced to endure the obsolete acid-etched zinc discs, but this was remedied, and 1900 also saw the introduction of paper labels on records.

Not everyone was happy about the success of the gramophone, or the quality of the tunes it was perpetuating. In January 1899 the society preacher Stopford Brooke wrote in his diary :

> Heard the gramophone—a vile concoction of the scientific people. Cannot they let us alone? Why will they reproduce the human voice, and if they do it, why should they choose music hall songs for reproduction? It is a revolting thing to listen to . . . It is an ingenious piece of work, but the voice that came out of it was like the voice of a skeleton—a weird, vile, uncanny,

76   A fairly late phonograph by Edison

monstrous thing! I hate it even more than I hate the telephone, and all its ramified iniquities.

The gramophone became incorporated into popular language, as we can see from an extract from a poem published in a weekly picture paper in 1901:

Sweetheart, you've heard your song and laugh
   Repeated true and clearly;
But won't you tell the gramophone
   'My Jim, I love you dearly'?

In Britain the quality of the music being put on record was higher than in America. Caruso was an early recruit; before his death in 1921 he had earned over £400,000 from the gramophone. Sir Landon Ronald, the conductor, was employed by the Gramophone Company as musical adviser, and he did much to counter the prejudice of musicians against the gramophone. In 1903 twelve-inch records were introduced, with a four minute playing time; it was not until 1948 with the coming of microgroove and long-playing records that the twelve-inch disc became obsolete. From 1900 to 1925 records revolved at between 74 and 82 revolutions a minute. Electrically powered turntables set the norm at 78 rpm.

The early years of this century saw further developments in the gramophone, especially in the size of the horn. The genteel thought this somewhat obscene, and in America in 1906 the gramophone was civilised by being incorporated into a piece of furniture, with the horn hidden away. This was the celebrated Victrola (Victor was one of the two principal companies in the United States). The phenomenal success of the Victrola, retailing at $200, had repercussions elsewhere. The high price of some records was accepted willingly (the price depended on the status of the performers—a twelve-inch single-sided disc could sell at $7). The phonograph-makers watched with envy; their share of the market was becoming smaller each year, though in 1913 there were still a million phonographs in circulation.

Many music lovers regretted the four minute limit imposed by records, and in 1904 Neophone records of plastic on cardboard were made which were twenty inches across, and could play for up to ten minutes. In sound and quality they did not compare with the records they tried to oust, and were withdrawn two years later.

77 *The graphophone, introduced in 1889. This model dates from 1897.*

In America the dance mania of 1914 shot the sales of gramophones sky-high, and in 1917 the Original Dixieland Jazz Band made a record that was to precipitate a new gramophone-buying public. In 1919 American gramophone manufacturers were making two million machines a year. It seemed as though the gramophone had reached the end of the road, incapable of further improvement. The Victrola of 1919 was almost identical with that of 1906.

The British were following the American lead but on a muted scale. As soon as the gramophone had achieved respectability they began disguising it, and in 1919 it was housed in splendid cabinets, and with names such as the Duncan Phyfe, the Aeolian Vocalion, the Deccalion, or the Oranola (in the form of a side table) the gramophone had truly arrived as a piece of furniture. Externally, there is no clue as to the mechanical contents, and the Duncan Phyfe, made in walnut with finely reeded legs, could be taken as a piece of superb Edwardian cabinet-making. Less commendable was the practice of tearing out the guts of Georgian furniture to house the gramophone.

The coming of the radio shook the gramophone industry to the core; the first tremors were felt in 1922, and to stave off possible effects the makers began introducing radiograms. At about the same time there was talk of electric recording; acoustic recording had never been more than an approximation to the real thing. With the radio threat it was necessary to update the gramophone, for it was evident that, even in the early days of radio, there was a standard of tonal quality that had never been achieved acoustically. To reproduce properly the range of sound engraved by the electrical recording process, it was necessary to have a horn at least nine feet long. This, thought the scoffers, was ridiculous. This difficulty was overcome by having the horn folded back upon itself, and in 1924 a prototype was made. The Orthophonic Victrola was presented to a great fanfare of publicity in 1925. About the same time a competing firm, Brunswick, produced the first all-electric gramophone. In England, the prospect of a horn nine feet long did not deter the experts. A horn folded on itself was considered a contradiction in terms. The trendy gramophones with monstrous curving horns so sought after today date from the mid-twenties, and were produced not out of the quaint desire to astonish, but to repro-

The phonographic "waits" of the future.

78   The phonograph and its use, anticipated in this cartoon of 1898.

duce as well as possible the full range of tone promised by electrical recording. In this these monsters succeeded, and even hi-fi enthusiasts can be startled by the fidelity of the gramophones of that period.

The introduction of the all-electric Electrola to counter the all-electric Brunswick was successful, and the merger of radio and gramophone in 1929 made the future rosy for the big names of both radio and gramophone. Edison was still on the side-lines, still clinging to his outdated phonograph; his promise to make a long-playing record of twenty minutes' duration had not been fulfilled. The new combine R.C.A. Victor were on the verge of marketing a $33\frac{1}{3}$ rpm record.

The slump in America almost destroyed the industry. Record sales dropped from 104 million in 1927 to 6 million in 1932; gramophone sales from 987,000 to 40,000. The backlash of the American slump hit Britain, which rationalised its gramophone interests in the merger of H.M.V. and Columbia into E.M.I. The industry revived in 1933-4 with the introduction of a record player that could be plugged into a radio, and given an extra spur in 1935 with 'high-fidelity' recording, a high-sounding title that meant little. The end of Prohibition saw the introduction of the juke-box.

In Britain, the pre-war years saw the massive production of

portable wind-up gramophones. It was the age of the weekend bungalow and the Sunday on the river. During the war, British-owned Decca pioneered new recording techniques, and their 'ffrr' (full frequency range recording) captured a large slice of the American quality market when they were exported there in 1946. But the gramophone was stationary. Superb recording was countered by the four minute time limit and the inherent crackle and hiss of shellac. In 1948 long-playing records made their entrance in America, though not until 1952 did Britain follow suit. Vinylite replaced shellac. 78 rpm gramophones, no matter how aristocratically they were housed, became obsolete.

To what extent are gramophones collectable? To those with an affection for pre-electrically recorded discs it is sacrilege to play them on modern players, even though these do have a reversible stylus that can play either 78 rpm or long-playing records. The essence of the modern record player is the light-weight pickup; old records were intended for a steel needle set in a heavy mounting, and for this there is no substitute.

It would be idle to maintain that the first gramophones with a direct hand drive are functional; amusement at having to sustain 78 rpm by ear is short-lived. Gramophones produced in the golden years of the 1920s are a different matter, for although lacking the mechanical finesse of cylinder musical boxes they are generally excellently made, and this is also true of the portable wind-up gramophones, made for picnics and associated rough handling. Single-speed gramophones made in the transitional period between 1948 and 1952 are perhaps the least interesting of them all, for bakelite is perhaps the most unattractive substance known to man. Before the introduction of modern plastics bakelite was used to an extraordinary extent.

Gramophones were renamed record players with the coming of the long-playing record and the introduction of three speeds. The record players of the early 1950s already seem archaic due to the design revolution effected by German and Japanese manufacturers and the impact of transistorisation.

It seems difficult to believe today that in the early 1930s the gramophone industry was in such parlous straits. It had one further competitor after radio—the tape recorder. Magnetic recording dates back to 1899 when the telegraphone was invented; an electromagnet created magnetic patterns in a strip of steel, depending on the varying electrical impulses. The inventor used an ordinary telephone transmitter to convert sound to electrical impulses, and these impulses activated an electromagnetic recording head. The telegraphone received a warm welcome at the Paris Exposition of 1900, and many saw it as likely to supersede both the phonograph and the gramophone. Had it been invented in America the snags might well have been conquered, but it was thought up in Denmark. The two disadvantages were a very low sound level, and an inefficient transmitter. The telegraphone was used in a small way for office dictation, and a few were exported to America, but it was ahead of its time. In the 1920s the Germans took an interest in it, and a scientist named Pfleumer developed a paper tape coated with iron oxide in place of steel wire or tape. The result was the magnetophone of 1935, ideal for office purposes but inadequate for the recording and playback of music.

During the 1939–45 war the combatant nations developed the tape recorder, the British favouring wire recording. In this research the Germans were leading, and when the Allies captured Radio Luxembourg in 1944 they captured perhaps the finest tape recorder made to that date, playing 14-inch reels of tape at 30 inches a second. With the knowledge gained from a close examination of this machine the Americans were soon making tape recorders that were a viable alternative to the gramophone. Production started in 1947, and but for the introduction of long-playing records it is highly probable that the tape recorder would have swept all before it

To the technically minded, the tape recorders of the inter-war period are as intriguing as the earliest phonographs, perhaps more so as they are technologically more interesting, though it is doubtful if there are many about today.

There has been a recent revival of interest in early radio sets. Much of this is due to the desire to get hold of components such as valves that are no longer available, for incorporation in other electrical apparatus, but some of the cabinets and cases have a period charm. The appearance of English wireless cabinets was, in its first phase, dependent on historical styles of furniture. Radios, like gramophones, were housed in Queen Anne, Georgian or Tudor cabinets, constructed by cabinetmakers nurtured in the fine traditions of Edwardian furniture.

About 1930 radio design wanted its modernity to be expressed in the cabinet, and flamboyant sound holes made their appearance. The Cubist movement in art had a short-lived effect on radio design in 1932, but pure cubism was soon lost when radio makers began using the current trends now known as art deco or the *style Odéon*, with Egyptian and sun-burst motifs. Two British radio makers, Ekco and Murphy, broke away from the practice of using staff designers; Murphy employed one of the leading furniture-makers of the day, Gordon Russell, to design austere fitness-for-purpose cabinets, and Ekco commissioned Serge Chermayeff to design something in the multi-purpose plastic of the age, bakelite, invented by Dr Baekeland in the United States in 1908 and introduced to Britain 1921–2. In 1937 Dr Pevsner wrote of the Ekco bakelite radio cabinet: 'The shape of it was something completely new, nothing comparable existed, either in England or abroad. It was the result of a careful study of function and a genuinely artistic imagination.' In 1934 Ekco called in Wells Coates, who approached the problem 'as one of designing a piece of modern machinery'. Whereas the Chermayeff design appears nostalgically of the period, Coates's cabinet is much more interesting—a squat cylinder with the dial set round the periphery of one of the flat planes, the speaker in the middle, and the three knobs at the bottom. The cylinder is set on two stumpy legs. Cabinets of this kind certainly fit into homes being refurbished in the style of the 1930s; and why should they contain a radio?

The story of mechanical music is endlessly fascinating. In the nineteenth century, musical boxes and paper-roll organs were substitutes for the real thing; the musical box was orientated to the most popular instrument of the day, the piano, and the paper-roll organs related to church and concert ogans, in the same way that harmoniums were poor relations of church organs. Barrel-organs to some extent and most barrel-pianos were outdoor equivalents of indoor instruments.

The nearest approach to the real thing was to use genuine instruments, and have them play automatically, epitomised by three instruments of the 1890s, the Tanzbar mechanical accordion, the mechanised banjo, and the mechanised zither. But these were side issues. The greatest ingenuity involved was in the construction of a self-playing piano.

79 *The Berliner gramophone.*

Mechanical music of the nineteenth century centred on the pinned barrel, and for his *piano mécanique* Debain of Paris (died 1877) used not a pinned cylinder but a pinned series of planks, called planchettes. Debain supplied a second set of hammers working from above, set in motion by iron levers, the tops of which projected as 'beaks' through a four- or five-inch comb. Into this space five octaves of the keyboard were compressed. By turning a handle the planchettes were set in motion along a grooved plane, the pins acting on the ends of the levers, which in turn operated the hammers against the piano wires. By varying the length of the pins Debain could determine soft and loud, and accent.

It was ingenious and it worked, and several of these pianos made their way into Britain, but the attachment was expensive and limited in scope. It is not surprising that this innovation came from France, for they led the field in piano development. However, the first successful self-playing piano came not from France but from America, with the Angelus player-piano in 1897. Like the Organette this worked on the principle of a moving perforated paper roll. In 1898 the Pianola made its appearance, followed by the Apollo in 1900. Although the French had patented a player-piano activated by music rolls in 1863, the Americans were the first to bring the idea to fruition.

Between 1879 and 1902 a total of 55 patents had been issued in the United States, by which time it was possible to play music accurately and efficiently. Player-pianos were operated by foot-pedals; the principle of the player-piano rested upon the laws of atmospheric pressure. The main development was towards variation of touch. The term player-piano was eventually dropped, being replaced by reproducing piano. 'The ideal of all experimentation with the reproducing piano has been to strike any note at any time with any degree of force used by the actual pianist.' Touch could be analysed into weight and pressure. Correct tone colour obtained by a pianist by using the pedals was also desirable.

The first reproducing pianos to realise all the ideals were the mechanisms manufactured in Germany by M. Welte and Ludwig Hupfeld, but the Americans were not far behind with the Ampico, marketed by the American Piano Co, while the Aeolian Co put out the Duo-Art. The great piano-makers of the times such

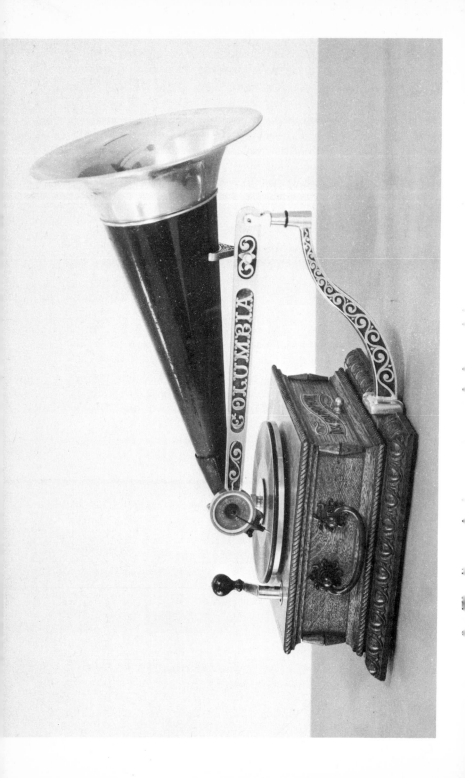

as Steinway, Weber, and Broadway welcomed these mechanisms for their pianos. By the use of levers and buttons an amateur could render any piece of piano music with the expression and technical ability of a master. Many famous pianists of the day 'recorded' music for the reproducing piano; the performer sat at an ordinary keyboard and played in his customary style, while, electrically connected with his piano and invisible to him, a recording machine carried an endless roll of paper on which the notes and the expression were marked as he played. The earlier player-pianos had their rolls made by technicians following the sheet-music; there was no attempt at interpretation.

The reproducing pianos of 1904 and after were marvels of mechanical skill, and when the foot-pumping was replaced by an electric motor, the reproducing piano became the most perfect example of mechanical music-makers. The production in America in the 1920s of exotics such as self-playing violins in sets of three enclosed in ornate cases and operated by electricity, and worth several thousand pounds, do not alter this.

The question one must ask when dealing with all forms of mechanical music is not whether the machine plays, but whether it plays well. The more sophisticated instruments should only be bought from reputable dealers, for a cylinder musical box that has a tinkling charm may have as many as half a dozen teeth missing from the comb or upwards of a hundred pins broken. Pins cannot be replaced on a cylinder except by a complete re-pinning operation, and teeth cost £5 each to replace (the expense is not in the replacing but in the tuning). Disc musical boxes are more substantial, and defects are more easily observed or heard. Much the same applies to gramophones or phonographs; one must not expect too much from the tone of early gramophones or phonographs.

Player-pianos can be found in great quantities, and fortunately there are thousands of music rolls available (the average price of piano rolls is 50p). A quality piano will usually have a superior player mechanism. The best player-pianos are usually very large. The player mechanism should be thoroughly checked, and the piano mechanism should be looked at, to make certain that the hammers are all there or are not too worn. The wires should be examined to see that they are not rusty; a small amount of rust can be removed with fine wire wool, but where the rust has really bitten into the wires restoration can be a fairly expensive project. If the notes clatter there is no cause for alarm; the felt pads beneath the keys can be replaced cheaply.

Because player-pianos are much more cumbersome than ordinary pianos they fetch less than one might expect. If one is buying a player-piano it is a good idea to find out how much carriage will cost. How does one tell a superior piano from a mediocre one? The sound, of course, is important, but this can be deceptive if one has no other piano to compare the sound with. An overstrung piano is always superior to what is known as a 'cottage' piano; in an overstrung piano the wires criss-cross instead of going straight up and down. Wooden-frame pianos should be avoided like the plague, though it is not often that one comes across a wooden-frame piano with a player mechanism. On a wooden-frame piano there is no metal bracing, and a further check can be made by lifting up one end of the piano; wooden-frame pianos are much lighter than metal-frame. The reason why they should be avoided is that they go very easily out of tune, especially where there is central-heating.

The maker's name should be noted. A good maker rarely makes a bad piano. Weber, Steck, Bechstein, Steinway, these can always be relied upon; Broadwood and Chappell, usually. An interesting tip is to look at the keys. In pianos ivory was put on the keys in two sections. If there is an almost imperceptible break along where the black notes come then the keys are made of ivory.

CHAPTER 9

DOMESTIC APPLIANCES

In no field does the adage that necessity is the mother of invention hold more true than in the domestic. Where there were servants to carry out the chores, domestic aids were not called for, except as novelties, and the basic set-up of a household unit did not vary much between, say, 1850 and 1890. The running of a house depended on a supply of domestic servants. In 1851 there were 908,000 servants in England, in 1871 1,503,000. In the whole of the United Kingdom in 1883 there were 1,951,000 servants; the home was the largest employer of labour. Servants' earnings amount to 68 million pounds a year, averaging out at £37 a head.

It is impossible to over-estimate the role of the servant in nineteenth-century life, and man and woman power accomplished all the duties that could have been taken over by mechanical means, from carpet-cleaning to keeping the house warm. The first high-pressure system of central heating, using hot water circulating in pipes, was perfected by A. M. Perkins as long ago as 1845, but it was considerably cheaper to have servants hauling coal from room to room. Coal had steadily dropped in price from 1820, when it was £2 13s 3d a ton to 1880 when it was 18s 4d a ton.

The need to introduce technology into the home was only apparent when there was no supply of servants, and thus America it was that provided the lead in most of the domestic gadgets we now take for granted. Immigrants from Europe accustomed to having servants were either forced to do the menial work themselves or to invent something that would do it for them. The skills were not lacking, and there was ample incentive.

Many of the ideas that had been in the air for some time were impeded by the absence of a small power unit, and it was not until the coming of electricity, in particular mains electricity, that some ideas could be brought to fruition. The home generator was the only means of manufacturing electricity until the early 1900s; the first turbine generator made its appearance in 1903, and in 1910 came the first hydro-electric generator.

America was mechanised from the start. The size of the country, the sparse population, the lack of labour and the high wages attendant thereon, all these made necessary the search for a substitute for man power. It was frequently a question of survival rather than comfort. The transportation of meat and fruit across vast distances through widely diverse climatic conditions made the invention of a viable refrigeration system on the railways essential. Harvesting thousands of acres with an inadequate labour force made the development of a mechanical substitute for man's labour imperative. As early as 1836 the combine harvester was introduced on the prairies; this would thresh, clean, and bag the wheat, and although such an invention would have been useful to the European farmer it was not evolved where there was peasant labour to do the job. Where there were vast distances to overcome, and when communications were unreliable and erratic, such mechanical substitutes for man had to be simple and reliable, and from this demand there arose the technology of interchangeable parts, which led naturally into the theory of mass-production. An American farmer who bought a combine harvester would therefore not only purchase the machine itself but a wide variety of spares that he himself could fit when the original broke or wore out.

Mass-production is the key to the American dream, and is the reason why the country achieved such a lead not only in the manufacture of domestic objects but in typewriters and sewing machines. Not that all the kudos can be laid at the door of the Americans. As early as 1839 J. G. Bodmer organised an assembly line at a Manchester machine-tool factory, assembly line methods were used in grain milling in 1783 and in Portsmouth in 1833 a biscuit manufactory used rollers operated by steam power to accelerate production. None of these techniques was taken up by other manufacturers in Britain in any kind of systematic manner, and even in cotton manufacture, the first industry to be thoroughly mechanised, correlation of processes was never as efficient as it could have been.

Pioneers of mass-production methods in America were Eli Whitney (1765–1825), the inventor of the cotton gin (1794), and Simeon North, a pistol-maker. At his arms factory in Whitney-ville, Whitney brought modern work-study to bear on gun manu-facture, and by the early 1800s was involved in the theory of interchangeable parts.

It is impossible to stress too much the importance of mass-production in the evolution of mechanical domestic appliances, for only by this method could prices be brought down to a real-istic level, only by the prospect of large sales could considerable development costs be approved. A good example of mass-pro-duction techniques brought to a fine pitch is furnished by the Waterbury watch factory, operating in the 1880s, which employed 380 people and made 1500 watches a day. Although there were 500 operations, the working parts of the watch had been brought down to 57. The result was that this watch could be sold very cheaply, not only in the United States but in England, and the London agency of the factory in Holborn Viaduct sold 202,000 watches in 1886, more than a third of the factory's output.

Two factors strike one in examining mechanical domestic apparatus produced in the United States in the last century— their early dates, and their massive sales. As early as 1851 a German observer, Lothar Bucher, noted that 'American domestic equipment breathes the spirit of comfort and fitness for purpose'. When British manufacturers were attempting to promote hum-drum equipment in terms of something else, when their stoves were disguised as pieces of furniture and iron and steel were tortured into shapes that had been relevant for wood, American entries to the Great Exhibition brought a no-nonsense air that affronted rather than convinced.

The kitchen offered the greatest scope for mechanical innova-tion. One of the most irksome and time-consuming jobs was laundering. In 1780, British Patent 1269 had projected a 'machine called a laundry for washing and pressing apparel' but

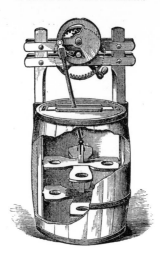

*81   This dairy produce maker of 1879 worked on the same principle
as the washing machine of its day—rotating blades, simple
gearing, and a hand crank.*

this was not proceeded with when there were laundrywomen to
do the same thing by hand. In the 1840s the French produced a
washing machine, which again was still-born, but in 1851 the
American James T. King invented a cylindrical washing
machine, in which the basic principle was a revolving perforated
inner cylinder. This, the first commercial washing machine, was
horizontal, and was relatively ineffective, for only one of the
factors involved in laundering was used—movement, but not
agitation. In the 1860s the familiar vertical cylinder washing
machine was evolved, with an agitator. The two movements,
rotation and agitation, were carried out by means of a foot
treadle or a handle. At about the same time it was realised that
this washing machine would serve also as a spin-drier, using
centrifugal force, and this relatively sophisticated machine led
naturally in 1878 to a two-speed single-tub washer/drier. It was
an early example of simple gearing being used for the service of
the housewife. No further refinements could be used until the
advent of a compact power unit, the electric motor, but notwithstanding the limitations of a hand-powered washing machine
sales reached $1,000,000 in 1870 and $3,700,000 in 1900. That
there was a market for washing machines was soon realised by

American manufacturers—in 1873 there were 2000 US patents relating to them. The key date in washing machine development was 1869 when the four-bladed rotor was introduced.

The mechanical dish washer almost qualifies as a genuine antique. First made in 1865, the principle was the same as that of the washing machine—agitation of water, provided by metal blades at the bottom of the tub. US Patent 51,000 describes the method in full: 'After the water has been thrown outward among the dishes it will flow back again to the centre. The plates are made to occupy a position tangential to permit the water to be thrown between them'. The dishes were placed in a wire frame. A good deal of thought went into the design of this dish washer, including a metal ring fixed to the rim of the tub 'for the purpose of preventing the water from dashing against the underside of the cover and working out around the edges of the tub'. The dish washer was worked by means of a horizontal handle operating a central axle driving the blades without further gearing. If a distinction has to be made between domestic equipment and domestic gadgets, there is no question that the dish washer belongs to the latter group. Nevertheless, the premature appearance of such items as the dish washer has made them more easily acceptable to the American housewife than the British, and although the dish washer does not occupy the same position in the American kitchen as the refrigerator or the washing machine it is not the rarity it still is in Britain.

An interesting example of the manner in which novel objects are more widely accepted in the American house than in Britain is furnished by the electric clock. Though introduced in 1927 the electric clock was almost unknown in the British home before the war, but in the United States in 1930 nine out of ten American homes boasted at least one electric clock.

The use of the electric motor revolutionised domestic equipment, and although Nicola Tesla in combination with the Westinghouse Corporation had produced a 1/6th horsepower electric motor for driving a fan in 1889, many propositions had to wait for mains electricity before their full potential was realised. This was especially true of the vacuum cleaner, which, like the dish washer and the washing machine, has an ancestry stretching back into the nineteenth century.

Foreigners to Britain invariably marvelled at the contrast

82 This type of carpet sweeper was introduced into Britain in the middle of the 1880s and in essence remains unchanged today.

between the dingy exteriors of London houses and their bright sparkling interiors, but this cleanliness, often more apparent than real, was only accomplished by hard work. Carpets, the most prominent of harbourers of dirt, were taken up and beaten or shaken, but these methods were not always effective, and the Victorian practice of laying many small rugs and carpets over large carpets which were tacked down was a recognition of the fact that room-sized carpets were not amenable to cleaning. The difficulty of keeping carpets clean was also responsible for the widespread use of linoleum (patented by F. Walton in 1860) by all social classes.

Any equipment that helped to clean carpets was therefore sure of a hearty welcome. The simplest of such mechanical aids was the carpet sweeper with a revolving brush, patented in 1858, which derived from a street sweeper of 1840 involving a circular broom and an endless chain. A bellows carpet sweeper was introduced in 1860, and about the same time a sweeper was introduced in which the wheels drove a four-bladed fan.

The snag about cleaners that blew air was that the dust removed would simply settle elsewhere, and in 1869 the logical principle was grasped—a suction machine, but although various patents were taken out no one stumbled on the complete answer. The most profitable line of investigation appeared to be compressed air.

About 1900 it was clear that there was going to be a breakthrough shortly, and in 1901 an American inventor rented the Empire Music Hall for a demonstration of his cleaner, a box about a foot square with a bag on top. The box was pushed over the carpet and two jets of compressed air were blown at the carpet; the theory was that the surface beneath the carpet would deflect the dust into the box. In the audience was an engineer, H. Cecil Booth. How could the cleaner do cushions and upholstery where there was no surface beneath the object to deflect the dust back? Suction was the answer, and Booth tried the experiment of sucking with his mouth against the back of a plush seat in a restaurant in Victoria Street, and although he nearly choked he was convinced that he could make such a cleaner and, offered financial help, he took out a patent (British Patent 17,433) in which the term 'vacuum cleaner' was used for the first time.

Two things were clear—the cleaning machine had to be mobile, and had to remove dust not only from the surface but from the innermost body of the carpet or material, a task that had foiled every cleaner so far. The 'vacuum cleaner' was therefore mounted on a carriage, and, the idea that a vacuum cleaner would be individually owned being in the future, the Vacuum Cleaner Company was founded. One of the firm's first jobs was to go over the carpets of a large London store; two machines were used, and half a ton of dust was extracted in one night. Another early contract was to clean the great blue Coronation carpet under the throne at Westminster Abbey in preparation for the Coronation of Edward VII in 1902, and this created such universal interest that the President of France, the Kaiser, and the Tsar of Russia all wanted vacuum cleaners.

The next stage was to invent a vacuum cleaner that would be portable within a building, and this was achieved by having a central power supply, with tubes on every floor, so that all that needed to be moved about the building was the cleaning mechanism. Several were installed in large London hotels (where a few are still working to this day), though the first was in Dickins and Jones store in Regent Street, soon followed by one in the House of Commons. They were operated by turbine fans, and were, and are, exceptionally efficient.

The portable vacuum cleaners hauled by horse from house to house used the petrol engine as its power source. Although

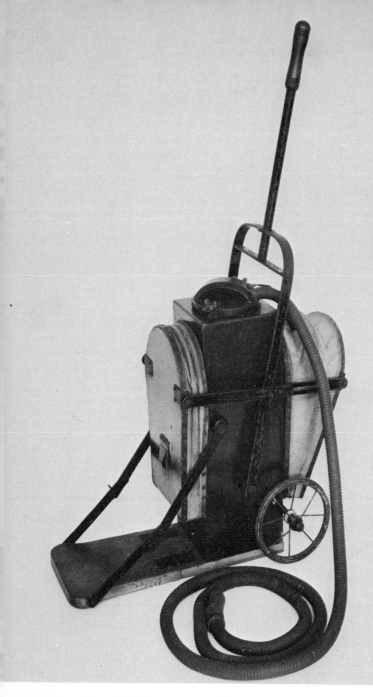

83 *A folding vacuum cleaner*

cumbersome these portable machines did all that the Vacuum Cleaner Company promised. They were too expensive for the average householder to hire, and it was not for several decades that the vacuum cleaner driven by electricity became a familiar feature in every home (many areas of the country were without electricity until quite recent times).

Although the lead was given by Britain, the Americans were the first to market the vacuum cleaner in quantity, and the ubiquitous Hoover was introduced, soon to dominate the field both in America and Britain and to furnish a handy synonym for vacuum cleaner. For a short time there was competition from water-driven cleaners; the 'Water Witch' was operated directly off a tap, and its advertising slogan of 'It never wears out, will last as long as your building' was countered by Hoover in 1909 with 'The motor will outlast the house you live in'.

By 1920 the vacuum cleaner had reached a pitch of perfection, and the models of that period are not a whit different from those manufactured today, and although strictly speaking they belong to a period that is at present being heavily collected (the early days of art deco) they do not have the appearance of bygone artifacts.

Though the refrigerator hardly comes into the category of a collectable item, its place in the history of domestic equipment is undeniable, and refrigerators of the inter-war period certainly have a curio value. Mechanical refrigeration was postulated by Michael Faraday in 1823. He observed that ammonia heated in a U-tube would recondense in the other limb, and left to itself would re-evaporate producing intense cold. The observation was not followed up, though when the first practical ice machine of Ferdinand Carré was invented in 1860, and shown at the Great Exhibition of 1862 (larger than the more famous one of 1851 but less well known) the same principle was adopted. This historically important machine was essentially a boiler three-parts filled with ammonia, standing in a stove, plus a vessel with double walls immersed in cold water, and although this might be termed the first refrigerator it was deemed by the exhibition visitors a marvel rather than a useful invention. The ice-house still held sway throughout the civilised world, and it was not until 1916 that refrigerators began to be produced in any quantity. In 1923 there were only 20,000 refrigerators in the

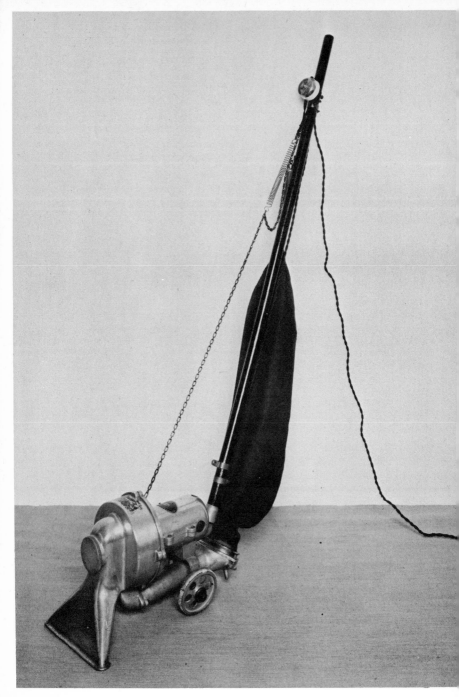

84 *An early electrical vacuum cleaner, called the Magic. To its users, it must have seemed so.*

United States. But in 1925 quick-frozen food was patented by Clarence Birdseye, and the refrigerator and frozen food technologies ran hand in hand, until by 1933 there were nearly a million refrigerators in the United States.

The discovery by Faraday of the properties of ammonia, the air machine of Dr A. Kirk (1862) described by him in a paper on the 'Mechanical Production of Cold', the first cargo of mutton brought from Australia in a refrigerated ship in 1879—it would seem that an inordinate amount of time was allowed to elapse before the perfection of an adequate domestic refrigerator. Whereas the electric-powered vacuum cleaner was born fully developed, the first domestic refrigerators of between 1916 and 1920 were hardly more than the old nineteenth-century ice boxes used by sportsmen to preserve game. The casing was of dark wood, and it was only later that refrigerators were streamlined and integrated into the modern kitchen. One of the chief reasons for the late arrival of the domestic refrigerator was the inability to provide an efficient thermostatic control.

Whereas refrigerators, washing machines and vacuum cleaners were not brought to perfection until the arrival of the electric motor, other domestic items were working perfectly well powered by other means. The cast-iron range, which was sometimes so intricate and ornamented as to qualify as a fine art piece, was in theory superseded by Graham's Gas Range exhibited at the Great Exhibition of 1851, but this indication of future trends was ignored in favour of Jeakes's 'stove somewhat of Romanesque or Byzantine taste' also shown in 1851. Gas took a long time to be accepted as a heating rather than a lighting agent, though in 1879 an English firm exhibited over 300 appliances in which gas was used for other purposes than lighting. Some of these appliances certainly fit into the category of domestic antiques though they were useless and ornamental; the gas iron, connected by a rubber tube to the gas jet, was unquestionably dangerous as well.

For even those most captivated by electricity, cooking by this method was considered an eccentric fantasy. In 1887 an electric saucepan, heated by current from a battery, was invented. Food cooked in it was said to have an 'electric flavour', though this did not deter organisers of an electrical fair held at the Crystal Palace in 1892, which exhibited electricity in the service of the

85   An ironing machine worked by gas dating from the 1880s.

culinary arts. An electrically cooked banquet in honour of the Lord Mayor of London in 1895 did a little to promote the new medium, but further major technical advances were halted by the absence of a serviceable thermostat. In design electric stoves followed the pattern of gas stoves, which were themselves influenced by cast-iron cooking-ranges, and the box on four legs remained the pattern until the early 1930s.

By 1920, electrically-powered domestic appliances were commonplace, and there was encouragement to use them by the introduction of a different rate by the electricity suppliers—the charge for electricity used for appliances was only a third of that used for electric light. A summary of uses of electricity in the home was compiled by a writer in *Our Homes and Gardens*—cooking, cleaning, ventilation, washing and wringing of clothes, ironing, peeling potatoes and mincing food, bed warming, medical treatment, playing the piano or driving a gramophone.

The installation of power points meant that the electricity user was not dependent on one power source, and the apparatus for cooking included electric toasters, hot-plates apart from a cooker, electrical kettles, electrically-heated covers for putting over meat

86 *The Apollo electric fire of 1904.*

87 *Electrical domestic gadgetry of the 1920s was anticipated by this 1895 cartoon lampooning mechanical gadgets.*

and puddings, portable electric rings, and small immersion heaters for warming liquids. There was also the advent of electric coffee mills and percolators, and 'electric egg-cookers' in which the egg was steamed instead of boiled. This was considered very handy for broken eggs. The electric teapot was one idea that never caught on (the tea was put in a wire container and lowered into the teapot) and the 'electric light bath' sounds as perilous as the gas iron. The new-fangled bedwarmers have been more recently named electric blankets, though there was an aluminium bedwarmer covered with felt in the shape of an old-fashioned hot-water bottle which did not survive very long. In the bathroom and bedroom there were shaving-water heaters, hair dryers, curling tong heaters, electric combs, electric vibrators for face and body massage, and illuminated mirrors, while a husband, bemused by all this apparatus, could 'light his cigar from the electric lighter fixed on the wall, though, unfortunately, no electric pipe-lighter has yet been evolved'.

Whether or not these merit consideration as antiques, they are certainly historically interesting and are unmistakably of their period. The designers had yet to refine the designs, and the toaster has a functional appearance with all the parts showing, not yet encased in steel or enamel. At the same time it has frivolities that hark back to the Victorians—ornamental roundels on the spring clip that holds the toast to the element, and small legs. The electric kettles of the period are ordinary kettles with the element inserted as if as an afterthought. Though not of cast-iron, the cooking apparatus from portable grillers to stoves have the appearance of it.

The present resurgence of interest in the art and the products of the 1920s and 1930s has not yet extended to the world of kitchen appliances, though it is only a short step to this. Such items as early electric razors, patented in 1913 by G. Appleyard of Halifax, are as worthy of attention as, for example, teapots of 1930 in the shape of a motor-car, or cigarette lighters in onyx and steel, items highly thought of by the cognoscenti.

Nineteenth-century domestic utensils already have their collectors; it was a logical step from the collection of wooden by-gones with an agricultural background. Cast-iron free standing stoves of the 1850s command £15 and upwards as they are ideal for conversion to electric light. Similarly such unique objects as the mechanical apple-parer of 1838, constructed on the principle of a lathe, or the patented egg-beater of 1857, madly intricate for such a simple job (refined in 1870 to a more rational shape) have an appeal, aesthetic, historical and technological, and much the same applies to non-starters such as the 1886 champagne stand in which a tube is inserted through the cork of an inverted champagne bottle and the liquid is drawn off in a tap (a second small tap is provided to allow air to enter to provide a smooth flow). Or the razor-stropping machine of the same year, in which it would take longer to set up the razor than strop by hand on a strap. Or the 'Agatha' candle guttering preventer, the precise details of which mystified the patentee.

Unquestionably in domestic appliances honours must go to the United States for her skill in rationalising the processes and for know-how in marketing, and even when the British were first on the scene, as they were in vacuum cleaners, they were at a loss as to how to promote the product. The British could invent, but they did not need to use. The innate conservatism of the race

A CLOCK THAT MAKES TEA!

Calls the sleeper at a given hour, automati cally lights spirit lamp, boils a pint of water for tea, shaving, and other purposes, pours it into a pot, extinguishes lamp, and finally rings second bell to signify all is ready. Invaluable to Ladies, Nurses, Professional and Business Men. It is strong, simple, and in no sense a toy. Prices 25s. to 70s. Postage in United Kingdom 1s. extra. With Foreign orders sufficient postage to cover 11 lb. Weight should be sent.

Please send for Illustrated Booklet, post free from

AUTOMATIC WATER BOILER CO.,
26a, Corporation St., Birmingham,
LONDON OFFICE AND SHOWROOM—

31, George Street, Hanover Square.

88   *Advertisement for an automatic tea-maker well ahead of its time.*

meant that such a beneficial apparatus as the gas stove was a long time being accepted; just as the 'electric flavour' was objected to when the first electric saucepan was used, so did the British middle and upper classes complain of food cooked by gas. The success of an invention or enterprise depended on whether it made the dominant classes more comfortable. At its simplest level, the spiral spring was speedily developed after it had been invented. But behind the scenes, in the unknown kitchens and amidst the masses of servants, the fact that mechanical appliances could increase the efficiency rate by 100% was simply not interesting.

It must not be supposed that only the British and the Americans were involved in the invention of domestic gadgets and utensils. The French and Germans, too, contributed towards the rationalisation of household chores (the man who invented the first refrigerator was French), but the social conditions were similar to those of Britain—a multitude of servants and the lack of necessity to make the servants' lot easier. The French had an inborn inferiority complex about things mechanical as the Queen made clear when she visited the Great Exhibition of 1851— 'there was much French machinery, which the French themselves fear they will not shine in'.

CHAPTER 10

GADGETS

BRITISH TECHNOLOGY, like British science, has always been on a stop-go basis, and the industrial promise hopefully foretold in the Great Exhibition of 1851 was not fulfilled. The reasons were various, and some credit must be given to the view that when Prince Albert died in 1861 technological incentive died too. Prince Albert came from a race and culture where patronage of science was a recognised necessity, and the gradual overhaul of Britain by Germany in the industrial stakes was due quite as much to the attitude of the rulers as to the innate ability of inventors and industrialists.

Although Queen Victoria rather surprisingly gave the go ahead to the cultivation of the telephone in Britain, she was by no means *au fait* with mechanical development. A simple example of the way a lead could be given or refused was furnished by the respective attitudes of Queen Victoria and her son Edward VII to the motor-car. Queen Victoria declared: 'I hope you will never allow any of those horrible machines to be used in my stables. I am told that they smell exceedingly nasty, and are very shaky and disagreeable conveyances altogether.' On the other hand, Edward VII was driving as early as 1899, and gave encouragement to the small and insignificant motor industry in Britain, giving the prefix Royal to the Automobile Club in 1907, and using a Daimler in preference to more highly rated foreign cars. In 1904 most cars in Britain were foreign, but by 1910 a good proportion of the 100,000 cars in use in Britain were made in England, and the number of motor taxis in London had risen from 2 to 6300. Without the personal lead given to motoring by Edward VII it is doubtful whether Britain could have wrested the home market from the French and the Germans.

By the middle of the nineteenth century, the tradition of the gentleman/scientist was dying out. In the eighteenth century Lord Shelburne had supported important work by Joseph Priestley (1733–1804), and discoveries of great moment were made by Cavendish in his own private laboratory in Clapham. It was not a fad for literary men to indulge in science—Dr Johnson reserved an attic for the pursuit of 'chymistry'.

The decline of science was noted by Charles Babbage, pioneer of the calculating machine, who wrote *Observations on the Decline of Science* in 1830, and the formation of the British Association for the Advancement of Science in 1831 hardly changed the general opinion that science was no longer the province of a man of parts. It is not surprising therefore that the men most involved in technological advancement were not scientists but engineers, many of them of a lowly social status, nor that such men were not backed and promoted.

Scorned and disillusioned, many inventors gave up when they were three-quarters of the way through a project. Their work was not wanted, and only when a device was *essential* for the efficient working of industry was it followed through. This happened with the telegraph, needed for the expanding railway system. Lack of imagination damned the telephone, far more important to economic and social life than the telegraph it supplanted, and aids to living and industry such as the typewriter, the sewing machine, the washing machine or the vacuum cleaner were contemptuously dismissed. Why invent a machine to supplant the clerk, available in large quantities and at minimal salaries, the needlewoman, who would die of starvation deprived of her work, the charwoman, willing to work for a few coppers a day, or the washerwoman, who operated on much the same wage scale as the char?

In places where labour was short and wages extremely high, but where there was the enterprise so notably lacking in Britain, then these labour-saving machines were eagerly welcomed, and even when initial backers dropped out there was always someone else to put money into what the British thought were scatterbrained ideas. The Americans went beyond the prototype stage, where British—and European—innovations had stuck, and were far-sighted enough to think in terms of quantity production.

Mass-production could have been used in Britain, for British toolmaking has always been pre-eminent, but it was not considered necessary, and although a primitive production flow technique was used at the Enfield arms factory it was never so streamlined as Colt of America, pioneers in mass-production. The preoccupation in Britain and Europe with the individually made object led to the massive intrusion of the American motor-car industry in the early years of this century. The Ford company could manufacture and ship their cars considerably cheaper than European companies clinging to the concept of the superior handmade object.

The regular pattern of invention and development in Europe, completion and marketing in America, held true not only of mechanical items mentioned in previous pages, but of simple devices such as the safety razor. This was invented in Sheffield in 1875, by William Gillette, a cork salesman; it took advantage of the ability of modern steel rolling to produce strip of one thousandth of an inch, but it was not exploited until 1895. It took a considerable time for British manufacturers to realise the possibilities of cheaper steel, made through the Siemens open hearth process or the Bessemer converter. When it was used for structural purposes it was overlaid with stone, to hide the stark reality, as in Tower Bridge (1894).

Mechanical design was bedevilled by the desire to present an object in terms of something else, or to make an object appear to be made of another substance. Between 1837 and 1846 thirty-five patents were issued in Britain for 'the coating and covering of non-metallic bodies . . . coating surfaces made of wrought iron which may be used in substitution of japanning and other modes now in use . . . mastic or cement which may be also applied as an artificial stone for covering metals'.

Mechanical antiques may be dated by the manner in which articles were disguised. Electroplating, invented in 1837, gave the impression but not the reality of silver, and electroplate was widely used as a decorative extra on mechanical objects. Even papier-mâché, used extensively throughout the early Victorian period for trays and small furniture, blotters and 'ceremonial ware', was pushed into the arena as substitute material for mechanical objects (the nearest approach to success was the papier-mâché piano).

Certain objects were made in a totally unsuitable medium for prestige reasons. A match box of one specific design could be made out of ormolu (4 guineas), bronze (30 shillings—£1.50) or Parian china (4 shillings—20p). Because marble clocks carried status. slate was treated to look like marble, and the most trivial articles were lavished with mother of pearl because it looked expensive (though it was not). Mechanisation throughout the nineteenth century meant that there were more articles stamped out, pressed out, and punched out, and there was much greater facility in making matrices or dies. The sordid ancestry of such articles necessitated their being worked on to present the image of being handmade. Sometimes the disguise can be very good, and humble cast iron can occasionally only reveal itself by weight. It was considered the highest duty of the mechanic and the engineer to present his product in the guise of an art object, and in 1832 Lord Ashley, later the Earl of Shaftesbury, considered that the erection of a gallery containing 'approved' mechanical devices 'would be extremely beneficial for artists and mechanics to resort to, and he had reason for believing that it would be frequented by the industrious classes, instead of resorting to ale-houses, as at present'.

It was the custom to be ashamed of bare mechanism. Many Victorians wished to turn the clock back, to capture the idyllic age before the Industrial Revolution. The most powerful of the voices was that of John Ruskin : 'The great mechanical impulses of the age, of which most of us are so proud, are a mere passing fever, half speculative, half childish.' The mere idea that mechanisms could be beautiful would have driven Ruskin to apoplexy; he would have been totally bemused by the philosophy of Dr Gropius, of Bauhaus fame :

> The transformation from manual to machine production so preoccupied humanity for a century that instead of pressing forward to tackle the real problems of design, men were long content with borrowed styles and conventional decorations. (1934)

Gropius made a distinction between handicrafts and industry. Such a distinction was difficult to make in the transitional stage between the one-off and the mass-produced object. The confusion is exemplified by the words of one of the covering patents already noted—the mastic or cement which could be applied to cover metals, a confusion compounded of shame and the inability to think things through. Applied cement to cover metal is one step further than applied ornament. Applied ornament was used to disguise or prettify. The flowers painted on to the chassis of the Victorian sewing machines are attractive and acceptable; sewing machines in the form of horses and cupids are interesting and desirable from the collector's point of view, but they scarcely enhance the sewing machine as a sewing machine. Even machine-tools were served up to look like baroque extravaganzas.

Fitness for purpose unquestionably makes for efficiency but not necessarily for collectability. The bentwood and bentmetal chairs and tables of the 1920s and 1930s, lauded by a generation of theorists and professors of design, sometimes because they were so deceptively expensive, are hardly more collectable today than plastic serviettes culled last week from a roadhouse. The light fittings and radiators cooed over by the Bauhaus set are of considerably less interest than the brash commercial bric-à-brac of 1930 cinemas and hotels. The furniture of such art and craft designers as Ernest Gimson, the doyen of Edwardian functionalism, is desirable not because of its fitness for purpose, but for its oddness and eccentricity, and the same applies to the products of Roger Fry and the Omega Workshop, ostensibly functional and in the height of *avant garde* taste.

Nineteenth-century mechanical gadgetry appeals to us in much the same way. The smoker's companion was intended as a perfectly valid useful thing; to us it seems to do a job that was never necessary, and a spiral spring for poking out the end of a pipe encased in a heavy pocket-size tin does not seem a desirable substitute for a length of wire covered with material (the common pipe-cleaner). The machine for tying up bouquets of flowers, with levers to raise and lower the flowers, and push them this way or that, plus an intricate arrangement for winding the wire around the bouquet, is gadgetry for the sake of gadgetry. It took as long to set the machine up as to tie the flowers by hand. There were hopelessly inefficient gadgets, such as the Eclipse copying apparatus, which claimed 'clear black copies equal to lithography' churned out at the rate of 500 per hour, or the Indis-

89 A patent gumming machine of 1886.

pensable gumming machine of 1886 which deposited adhesive
everywhere. The Agatha candle guttering preventer was of more
danger than utility, and the addition pencil may have been of
service to a five-year-old child but hardly to the commercial
class it was designed for; it comprised a small plunger and a
simple mechanism set in the body of a pencil that added in
single digits.

Inventors were kept at their trade by the knowledge that
patents could provide enormous sums of money. The so-called
stylographic pen and a pen for shading in different colours were
reputed to yield £40,000 per annum, and a simple device such
as an inverted glass bell to hang over gas to prevent ceilings
being blackened was claimed to have made more than £100,000
for its inventor. Toys were ready money-spinners; a wooden ball
with an elastic attached made its patentee £10,000 a year, and
a firework called Pharaoh's Serpent was equally lucrative.

Among the inventions patented in 1889 was a ventilated dress
belt, an improved apparatus for propelling balloons, a device
for opening and shutting the doors of hansom cabs, automatic
supports superseding crutches, and a magnetic gadget 'in connection with locks for boxes'. A more extravagant invention comes
from 1872, a walking stick that would convert into a pair of
steps in the event of the owner meeting a mad dog.

Yet there were patents that anticipated. 'Atomised solid fuel
for generating heat' may be dismissed as misleading, but 'litho-
graphing by means of sand blasting' has been widely used in the
manufacture of decorative mirrors, especially in the 1930s.

SCIENCE IN THE POULTRY-YARD.

No, gentle reader, you are mistaken. This is simply a little stamping attachment invented by our old friend Rott, by which a person purchasing eggs can tell to the minute when they were laid. The upper picture shows the attachment, and the lower the result.

90   An advertisement of 1886.

Illuminated clocks subsequently came into being, and filter-tipped cigarettes were anticipated in 1889 by W. Sharp of London.

There were a number of variations on penny-in-the-slot machines, already set up to deliver chocolate, sweets, matches and cigars. One of the most ingenius was the automatic photographic apparatus invented by M. Enjalbert of Paris, in which the extremely messy wet collodion process was used. One suspects that this apparatus received short shrift from a meddling public, but the penny-in-the-slot machines are highly collectable and have been for some years.

It would be idle to pretend that the patented combination toothbrush and nail trimmer, glass tombstones, pickle fork holders (will fit any size jar), hat ventilators, and novelty perambulators guaranteed to make any child sick within twenty seconds set the world on fire, but they do illustrate the ingenuity, if misplaced, shown by amateur inventors in an alien world. The extreme inconsequence of late Victorian gadgetry reflects the indifference of the establishment. There was little systematic backing of mechanisation.

It was very different in Germany. German banks, based on the French *Crédit Mobilier*, were organised to finance industry. British manufacturers were loaded with obsolete machinery, impossible to replace because of the lack of available capital. Science and industry were locked in logic-tight compartments, whereas in France and Germany polytechnics and their equivalents (dating from 1794) promoted cohesion. The twin institutes in Britain that could have provided a lead wilfully ignored science; Oxford and Cambridge, and the public schools, were concerned with largely useless and obsolete subjects. Even mathematics was a poor relation. The bias against industry and trade arose from the belief that money and mechanical labour were degrading. In the last quarter of the nineteenth century, only two major inventions can be credited to Britain—Parson's steam turbine and the Dunlop pneumatic tyre.

While British inventors were involved in the production of trivia and what has been described as 'snail-watching' they were overtaken by America and European countries, including Scandinavia, where Alfred Nobel was working. The possibility of using the internal combustion engine was not seen, even though an Austrian, Siegfried Marcus, made such a machine in 1864, and when Karl Benz constructed a three-wheeled vehicle driven by a horizontal single-cylinder petrol engine in 1885, and Gottlieb Daimler made the first motorcycle in 1886, these portents were disregarded. The one man in Britain who could have entered the arena at this stage was Edward Butler, who patented in 1887 his 'petrol cycle' (first use of the term petrol), but he was deterred from continuing his experiments by the inane restrictions against the use of road-vehicles in force in Britain. Notwithstanding the pioneer work of F. W. Lanchester and Austin in the 1890s, continental pre-eminence in motor cars persisted into the twentieth century.

The development of the internal combustion engine was handi-

91   An ironical comment on the penny-in-the-slot craze.

capped by the pursuit of the electric road-carriage. An electric tricycle was made in 1882, but the first electric vehicle to be more or less successful did not arrive until 1893. In 1897 belief in the electric vehicle was sufficiently strong for the London Electric Cab Company to begin a service of electric cabs, and when the company closed down two years later they had acquired a fleet of thirty-six vehicles.

Notwithstanding the cumbersome array of batteries needed before the advent of generating stations, the chimera of electricity was relentlessly pursued. The constant need to charge the batteries made, and has continued to make, the electric vehicle a figurative non-starter. But in gadgets the small power source needed made electricity more than attractive. Electricity was a magic word. Dr Scott's Electric Hair Brush of 1883 and the Harness Electropathic Belt of 1891 did not even use electricity. Galvanic machines were no more efficacious; they derived from the earliest of electrical experiments in which amusement was derived from having an electric current run through one. The galvanic machines consisted of a battery, wires, and two handles;

by grasping the handles the user was put into circuit, and it was considered that the mild electric shock thus induced was therapeutic. Psychologically perhaps it was. Galvanic machines, and most electrical apparatus connected with medicine and healing, are splendidly finished and housed in handsome wooden cases. These, especially the more obscure models, turn up repeatedly in auctions and in junk shops for a pound or two, and deserve collection. From the number that are about it would seem that the advertisers' blurbs were well heeded.

Other Victorian electrical gadgetry includes the electric cigar lighter, evolved in the mid-1880s; directly anticipating the cigarette lighters that are fitted into the dashboards of cars. This product of the Westinghouse Electric Company was described in these terms:

> When the current is turned on, a bunch of bare wire loops at the end of the handle, which are contained in incombustible material, are heated to a red heat and the cigar is lighted like you would light it from a red hot coal.

Electric tailors' dummies were a short-lived fad. In one shop window stood a model of a policeman which tapped with its staff on the window. The jointed arm was worked from a battery.

By the end of the 1880s more and more attention was being focused on electricity as a lighting agent, and many enthusiastic inventors forsook electricity and returned to the mechanical, coming up with 'chicken hopples which walk the chicken right out of the garden when she tries to scratch, the bee moth excluder, which automatically shuts up all the beehives when the hens go to roost, and the side-hill annihilator, stilts to fit on the down-hill legs of a horse when he is ploughing along a side hill'.

The collecting of mechanical and electrical gadgets has a definite appeal, but in terms of antiques they can hardly be regarded as 'classic' items. Mechanical antiques may be divided into two main categories—those of historical importance, such as early electrical apparatus including pioneer telephones and telegraphs, early cameras and cinematographic equipment, and 'firsts' of everything; and items that fit into a series and can be related each to each, such as typewriters and sewing machines, or, venturing into a higher price bracket, musical boxes. There is no greater field for the specialist. Although mechanical toys,

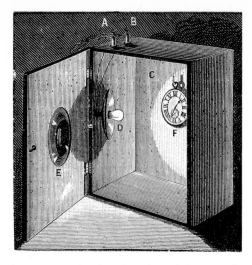

*92   An illuminated watch of 1885. A and B are wire terminals
leading to a battery, and E is a magnifying lens.*

for example, have been collected for some years there is considerable scope for the collector on a small budget. Mechanical
domestic apparatus, though perhaps lacking in intrinsic appeal,
has very few devotees. A representative collection of radio equipment could over the years be built up at a reasonable cost, and
phonographs and gramophones are decidedly underpriced in
view of their interest.

There is an attraction about antiques that work, whether it is
the smooth functioning of a cylinder musical box, the impressive
creaking of monster disc musical boxes, or the busy to-ing and
fro-ing of aged sewing machines. The great advantage of these
objects over modern electric and electronic apparatus is that one
can see how they work; it needs no mechanical genius to work
out the chain of events of even the most complex of mechanical
antiques, and it is worth-while travelling a long distance to see
automata, priced way out of the average person's reach, in action.

A good way to get acquainted with the various objects
included under the heading of mechanical antiques is to pay a
visit to the superb science museums, such as those at Cambridge,
Oxford, and Birmingham, and, of course, the Science Museum
in London. With a certain amount of observation and industry,

one will be surprised how often items in these museums are encountered at a reasonable price in antique and junk shops. And this cannot be said of many categories of antiques.

A SELECT BIBLIOGRAPHY

BOOKS

Anon	*Eighty Years of Progress in the United States*	1861
Appleyard, R.	*Pioneers of Electrical Communication*	1930
Baldwin, F. G. C.	*History of the Telephone in the United Kingdom*	1925
Beecher, C.	*The American Woman's Home*	1869
Bernal, J. D.	*Science and Industry in the Nineteenth Century*	1953
Buchner, A.	*Mechanical Musical Instruments*	1961
Butler, R. R.	*Scientific Discovery*	1947
Cardwell, D. S. L.	*Organisation of Science in England*	1957
Casson, H. N.	*History of the Telephone*	1917
Chapuis, A. and Droz, E	*Automata*	1958
Chapuis, A. and Gelis, E.	*Le Monde des Automates*	1928
Clapham, J. H.	*Economic History of Great Britain*	1933
Clark, D. K.	*The Exhibited Machinery of 1862*	1864
Cooper, G. R.	*The Invention of the Sewing Machine*	1968
Current, R. N.	*The Typewriter*	1954
Durham, J.	*Telegraphs in Victorian London*	1959
Dyer F. L. and Martin, T. C.	*Edison: His Life and Inventions*	1910

Eder, J. M.	*History of Photography*	1945
Ensor, R. C. K.	*England 1870–1914*	1936
Fahie, J. J.	*History of the Electric Telegraph*	1884
Ferguson, R. M.	*Electricity*	1866
Gaisberg, F. W.	*The Music Goes Round*	1942
Gelatt, R.	*The Fabulous Phonograph*	1956
Gernsheim, H. and A.	*History of Photography*	1955
Giedion, S.	*Mechanization Takes Command*	1948
Hall, A. R., Holmyard, E. J., Singer, C., and Williams, T. I.	*History of Technology Vols IV & V*	1958
Hillier, M.	*Dolls and Dollmakers*	1968
Hopwood, H. V.	*Living Pictures*	1899
Knight, E. H.	*American Mechanical Dictionary*	n.d.
Lardner, D.	*Electric Telegraph*	1855
Larsen, E.	*Ideas and Inventions*	1960
Passer, H. C.	*The Electrical Manufacturers 1875–1900*	1953
Pevsner, N.	*Industrial Art in England*	1937
Quennell, M. and C. H. B.	*History of Everyday Things in England*	1938
Quigley, M. (Jr)	*Magic Shadows*	1948
Read, H.	*Art and Industry*	1934
Richard, G. T.	*Typewriters*	1948
Routledge, R.	*Discoveries and Inventions of the Nineteenth Century*	1900
Sabine, R.	*History and Progress of the Electrical Telegraph*	1869
Smiles, S.	*Industrial Biography*	1879
Stambaugh, E.	*History of the Sewing Machine*	1872
Tallis, D.	*Musical Boxes*	1971
Tate, A. O.	*Edison's Open Door*	1938
Tunzelmann, G. W. de	*Electricity in Modern Life*	1889
Webb, G.	*The Cylinder Musical Box*	1968
Webb, G.	*The Disc Musical Box*	1971
Weller, C. E.	*Early History of the Typewriter*	1921
Wolf, A.	*History of Science, Technology and Philosophy in the Eighteenth Century*	1938
Woodward, L.	*The Age of Reform 1815–70*	1962

Articles are to be found in many periodicals of the time and in journals and magazines of today. The older ones are often available for reference in public libraries and it is sometimes possible to borrow, through the inter-library loan system, those which are not available locally. The following is a list of the periodicals where useful articles will be found.

All the Year Round, Cassell's Magazine, Cassell's Saturday Magazine, Chambers's Journal, Electrician, Encyclopaedia Britannica, Engineer, Engineering Review, English Mechanic, Graphic, Harmsworth's Magazine, Household Words, Illustrated London News, Illustrated Magazine of Art, Invention, Iron and Science, Journal of the Institute of Electrical Engineers, Journal of the Newcomen Society, Journal of the Royal Institute, Lady's Realm, Our Homes and Gardens, Pall Mall Magazine, Pearson's Magazine, Picture Magazine, Scientific American, Sewing Machine News, Strand Magazine.

INDEX